The 2nd Digital Revolution

Stephen J. Andriole
Villanova University, USA

IRM Press
Publisher of innovative scholarly and professional
information technology titles in the cyberage

Hershey • London • Melbourne • Singapore

Acquisitions Editor:	Mehdi Khosrow-Pour
Senior Managing Editor:	Jan Travers
Managing Editor:	Amanda Appicello
Development Editor:	Michele Rossi
Copy Editor:	Bernard Kieklak
Typesetter:	Cindy Consonery
Cover Design:	Lisa Tosheff
Printed at:	Yurchak Printing Inc.

Published in the United States of America by
    IRM Press (an imprint of Idea Group Inc.)
    701 E. Chocolate Avenue, Suite 200
    Hershey PA 17033-1240
    Tel: 717-533-8845
    Fax: 717-533-8661
    E-mail: cust@idea-group.com
    Web site: http://www.irm-press.com

and in the United Kingdom by
    IRM Press (an imprint of Idea Group Inc.)
    3 Henrietta Street
    Covent Garden
    London WC2E 8LU
    Tel: 44 20 7240 0856
    Fax: 44 20 7379 3313
    Web site: http://www.eurospan.co.uk

Copyright © 2005 by Idea Group Inc. All rights reserved. No part of this book may be reproduced in any form or by any means, electronic or mechanical, including photocopying, without written permission from the publisher.

    Library of Congress Cataloging-in-Publication Data

Andriole, Stephen J.
  The 2nd digital revolution / by Stephen J. Andriole.
    p. cm.
  Includes bibliographical references and index.
  Summary: "This book tells readers how technologies and business models are converging, and looks at technology and business holistically, arguing that it's no longer possible to think about business or technology without simultaneously thinking about the other"--Provided by publisher.
  ISBN 1-59140-801-6 (hc) -- ISBN 1-59140-598-X (sc) -- ISBN 1-59140-599-8 (ebook)
  1. Technological innovations--Management. 2. Business enterprises--Technological innovations. 3. Business planning. I. Title: Second digital revolution. II. Title.
  HD45.A644 2005
  338'.064--dc22
                                                    2004028470

British Cataloguing in Publication Data
A Cataloguing in Publication record for this book is available from the British Library.

All work contributed to this book is new, previously-unpublished material. The views expressed in this book are those of the authors, but not necessarily of the publisher.

The 2nd Digital Revolution

Table of Contents

Preface vi

Chapter I. Who's Here? 1
    Big Guns 3
    Bean Counters 5
    Techies 6
    Hype-Sters 7
    Protection 8
    Consigliore 8
    Other Chiefs 10
    Worker Bees 11
    Constituents 11
    Herding Cats? 13

**Chapter II. The Convergence Conversation –
We're Finally Ready** 15
    What Happened? 16
    Business Technology Convergence 28
    Early Takeaways 42
    Convergence Excellence 43
    Does This Make Any Sense? 44

**Chapter III. The Business Conversation – Where
We're Going** 53
    So What Do You Think? 54
    Inside/Outside Collaboration 58
    Supply Chain Planning And Management 60
    Customization And Personalization 64
    Real-Time Analytics And Optimization 68
    Automation 70

    Trust *74*
    Business Convergence Scenarios *75*
    So What Does Everyone Think? *79*

**Chapter IV. The Technology Conversation –
How The World Should Work** **84**
    What Do You Know? *85*
    Applications Integration And Interoperability *88*
    Data Integration *98*
    Pervasive Communications *103*
    Adaptive Infrastructures *111*
    Security And Privacy *116*
    Business Technology Convergence *121*
    Opinions? *122*

**Chapter V. The Turf Conversation –
Who Does What To Whom** **125**
    Watch Your Flanks *126*
    You Report To Who? *127*
    Organize – Because We Have To *128*
    The Special Case Of Innovation *137*
    Business Technology Organizational Convergence *141*
    Who Wants To Go First? *142*

**Chapter VI. The Management Conversation –
It Still Needs To Make Sense** **143**
    You Still Think So? *144*
    How Much Do You Know? *144*
    Variation's Your Enemy *152*
    Can We Really Do This? *158*
    Who Gets The Check? *167*
    Many Happy – And Miserable – Returns *178*
    The Business Of Business Technology Cases *180*
    Governance *188*
    Business Technology Management Convergence *214*
    Anyone Want To Talk About All This? *214*

**Chapter VII. The Tough Conversation –
It's Still (And Always) About People** **217**
    Who Are These People? *218*
    How's The Team – Really? *221*

Stay Sharp 229
They're Alive, They're Alive! 232
Keepers 235
Business Technology People Convergence 238
Why Is Everyone So Quiet? 247

Chapter VIII. There's Just One More Thing **250**
How'd You Do? 251
What's Not To Like About This Outcome 252

About the Author **254**

Index **258**

Preface

In the 1990s I sat in the office of a CFO of what was then a Fortune 100 company. He was not happy about the annual technology bill. Back then – and for decades before – technology was tactical. He said something about technology being his last unmanaged expense. I gave him a list of 10 things we should do to improve the cost-effectiveness of our technology investments. He told me to come back when there were only three things on the list. There was no appetite for long discussions about what was wrong – or right – with technology, and there was an expectation that technology expenses could be reduced by focusing on the top three problems.

The conversation was anything but strategic. It was about technology as real estate, or technology as furniture, something we had to have, a necessary evil, a cost incurred to support the transactions that made us money. The business itself was never discussed. It was almost as if technology existed independently of business models and processes.

Is this episode representative of the tension that still exists between "business" and "technology" in many companies, and the lack of synergism between business and technology initiatives? Unfortunately, yes. But in order to be as fair as possible, let's place such episodes in the context of the evolution of business technology. The 1970s, 80s and early 90s all constitute the 1st digital revolution, when enterprise computing and communications technologies were deployed within countless small, medium and large corporations. But since the mid-1990s, the capabilities and purpose of technology have evolved from the back office to the front office, from keeping the books to touching customers. The transition from "tactics" or "operations" to "strategy" marks the beginning of the 2nd digital revolution.

The early e-business initiatives – the Web sites we all had to have in the mid- to late-1990s – were the result of creative thinking about how to reach customers through the connectivity enabled by the Internet. These initiatives – embodied in e-business models like those offered by Amazon.com and eBay as well as just about every company on the planet that had some form of Web presence – represented first generation business technology synergism - the very same synergism this book argues is the new strategic imperative for the 21st century.

Synergism? Reaching customers 24 hours a day, seven days a week is only cost-effective through the use of low-cost access technology like the Internet. And reaching customers 24/7 is good business since it's likely to result in happier customers (who are also likely to buy more products). The 24/7 business goal cannot be cost-effectively achieved without Internet technology, just as Internet technology needs good business models to justify its (even cheap) existence. Synergism here results in a whole greater than the sum of the parts.

This book is about the macro trends that our research and practice indicate are gaining momentum. The business trend is **collaboration**, which means that companies are connecting with their employees, customers, suppliers and partners in new, continuous ways. The second trend is **technology integration**, which means that those systems we all use to keep the books, communicate with customers, and keep our employees informed are starting to share data and use each other's analyses to trigger collaboration. The intersection of these trends results in a powerful business technology synergism.

The key point is that business collaboration and technology integration are intertwined. Collaboration is increasingly enabled by technology and technology is increasingly integrated, and therefore capable of supporting collaborative business models that assume both process and data integration. Business technology is still "tactical" and "operational" – computers and networks still have to work – but it's also now very "strategic" – it can make or break a company, especially as business collaboration and technology integration trends accelerate and companies discover how to make existing business processes more efficient while they add new collaborative processes.

Business and technology are no longer disconnected, no longer part of a process that begins with questions about "business" and ends with decisions about "technology." The perspective here is that discussions about business or technology cannot occur without the other.

At the simplest level, this book is about how companies can increase the return on their technology investments. But it's not a book about how to calculate ROI. Instead, the book challenges executives and managers to think differently about the relationship between business and computing and communications technology – you know, the stuff you spend tons of money on year after year in the midst of suspicions about its contribution to profitability and growth.[1]

The book explains how the relationship works and how it should evolve, where it's been and where it's going. All of the arguments are based upon assumptions about how business is becoming more collaborative and how technology is integrating. Collaboration and integration are changing everything, but in the wake of the dot.com collapse and the decrease in capital technology spending that followed, we now find ourselves at a crossroads. We can continue to assume that technology spending swings with the larger capital market pendulum or we can look at it all differently.

The implications of the collaboration/integration perspective spill over into how we develop strategies and operate our companies. Two quick examples. "Customer relationship management" (CRM) is both a business model **and** a technology. The business objective of connecting and servicing customers, suppliers and partners is enabled by CRM systems sold by software vendors like Siebel Systems and SAP. Call center support is also enabled by technology, which provides 24/7 Internet access to frequently asked questions (FAQs) and, as customer satisfaction surveys indicate, produces happier customers. There's an additional advantage to Internet-based call centers: they permit companies to reduce the number of call center operators they have to hire. Both of these examples highlight the thin line between business and technology. Without a business model **and** technology, neither CRM or call center initiatives could succeed.

Which comes first, business or technology? There are people in your company always thinking about how to improve or extend business processes like CRM, call center management, marketing, up-selling and cross-selling. And there are also people always thinking about technology, about how to manage data, access information and keep everyone connected. The big change is in the innovation process. Where most new business ideas used to come primarily from the business side of your organization, new business ideas can now just as often come from the technology side – as well as from both – which, I'll argue here, will yield the most successful new ideas.

The business technology relationship today is the result of the evolution of technology, the result of technology's journey from hardware and software pieces that barely worked and seldom fit together to acceptable reliability, and business' consistent appetite for information that's efficient and cheap.

Business requirements – which were relatively stable during the last quarter of the 20th century – nevertheless outstripped technology's ability to deliver bullet proof performance. Technology vendors and business technology consultants exploited the lopsided evolution – not maliciously, but as business men and women who saw opportunities to sell a little, deploy a little, and then sell some more – even if what they were selling was half-baked. Many companies have struggled with technology, loving it and hating it along the way. By the late-1990s, things actually began to improve. Hardware and software became much more reliable, which freed managers from the daily "putting-out-the-brush-fires" drills that distracted them from strategic business technology planning. We were also appropriately distracted by the need to make sure our computers and networks kept working on January 1, 2000 and that we had "killer" e-business models to stake our claims to the new digital economy.

The disconnect between business questions and technology answers accelerated just when serious integration technologies began to appear, when technology vendors began to make their hardware and software more compatible, easier to inter-connect and even willing to accept data and analyses from competitors' systems.

Business models began to get collaborative conceptually when managers began to understand the value of monetizing "customer life cycles" through up-selling, cross-selling, personal service and other business models that followed customers through the stages of their personal and professional lives. These models always made sense, but were difficult to implement because the necessary business connections among the parties didn't exist and because our technology didn't integrate. Technology integration is enabling real collaboration, like the CRM and call center examples discussed briefly above.

Over the years, a lot of business creativity was stifled by our inability to keep computer operating systems, applications and networks from crashing. Now we have the luxury to think about collaborative business models, like supply chain planning, personalization, customization, and automation, among other ways to connect employees, suppliers, customers and partners. We'll talk a lot about these models in the book, models that have always been conceptually elegant, but are now enabled by digital technology that's integrated and reliable.

It's important to understand the period from 1995 to 2000 – in spite of our obsession with the Year 2000 problem and e-business – as the period that built the foundation upon which serious technology integration now rests, and the time that launched serious collaborative business models. In all of the turmoil, many of us missed the new integration technologies that enabled new collaborative business models. We'll focus here directly on these huge though relatively under-hyped trends.

If you're sitting inside of a company that buys and deploys business technology, you'll love this book. This is a buyer's book. It's about the business technology relationship as engineered by those who run all kinds of businesses. If you're a technology vendor, the book will explore the depths of the love/hate relationship that sometimes exists among you, the businesses that buy your products and the consultants who wrap them in lucrative services. The good news is that you're finally starting to embrace some common standards that make business technology convergence possible. The objective here is to get everyone talking candidly about what needs to happen next, about how business and technology can profitably converge. The savvy consultants who participate in these conversations will like what they hear.

The book is about how business and technology are now – and forever – inseparable, life-long, inter-dependent partners. It's about the maturity of the business technology relationship and how it can be exploited for competitive advantage, and it's about how to optimize the relationship by tweaking how we manage the acquisition, integration and support of business technology. It's ultimately about a little business, a little technology, and a whole lot of forward-thinking common sense.

Perspective

Age brings a few advantages. One of them is perspective. I could not have written this book five years ago. I just didn't know enough about how all the pieces fit together. Like a lot of "technologists," I rounded out my understanding of the relationship between business and technology relatively late in my professional life – or only after I'd seen business technology from several very different perspectives, some successful and some horrendous.

So here we are. The interplay between business and technology has evolved to the point where the conversations in this book can actually occur – and might actually make sense. Chances are you've spent a lot of time and money figuring out how to optimize the relationship between your business and computing and communications technology. When you add it all up, we're spending well over a trillion dollars a year on hardware, software and services.

Maybe you're spending millions or tens of millions. A few of you are spending billions each and every year on these bells and whistles. Well into the trenches, according to the Gartner Group, lots of us are spending around $4,500 per year, per user to support wireless personal digital assistants (PDAs). Yeah, that's right. Those cute little devices that your senior – and not so senior – managers play with all day are costing your company $4,500 per year ... each. Can you find the business case for these little darlings?

We're at a unique point in time when three things are absolutely true:

- Computing and communications technology is actually starting to work. The stuff is coming together in ways we couldn't imagine 10 years ago and had trouble describing even five years ago.
- Business models are morphing (partially because of technology and partially for other reasons) and they're morphing toward collaboration, supply chain integration, personalization and customization, among other connectivity-based models and processes.
- The inertia of old business technology management practices is still, however, driving most of our technology investment decisions, still driving us toward piecemeal applications, ill-conceived sourcing and staffing strategies, crazy organizational strategies, and metrics that measure the wrong things.

Since 2000, there have been a series of high profile challenges about the value of information technology, and whether or not IT is still really important. For example, Paul Strassmann (www.strassmann.com) has argued for years that investments in technology do not predict profitability or growth. More recently, longitudinal research reported by Joyce, Nohria and Roberson in *What (Really) Works* (Harper Business, 2003) reports "no correlation between a company's investment in technology and its total return to shareholders." All of these – and other – arguments are made in Nickolas Carr's now famous piece in the *Harvard Business Review* with the provocative title: "IT Doesn't Matter."[2] Carr is convinced that technology's strategic impact has run its course, that the technology playing field is now level.

Has Everyone Lost Their Minds?

Should we believe that computing and communications technology are frauds, that they bring very little to the competitive table, that the $1+ trillion a year that U.S. companies spend on hardware, software and technology services is somehow misspent?

Arguments that IT no longer matters are horribly flawed. In fact, we're confusing several healthy trends with what some see as declining influence. For example, there's no question that PCs, laptops and routers are commodities. Even some services – like legacy systems maintenance and data center management – have become commoditized. Are PCs, PDAs and servers "strategic"? Of course not. But if we botch the acquisition of these devices, or fail to adhere to sound management practices like standardization, they become tactical liabilities. Far from being irrelevant, they're actually tactically necessary and potentially dangerous.

Another misinterpreted trend is the increase in discipline used to acquire, deploy and manage technology. We're much more sophisticated about the use of business cases, total cost of ownership models, return on investment calculations, and project management best practices than we were a decade ago. Put another way, the acquisition and management of technology has become routine, no longer the high profile, *ad hoc* process it once was. Does this mean that it's no longer important? I'd argue that our ability to more skillfully acquire and manage IT is an indicator of maturity, not unimportance.

Another trend that seems to confuse the technology-doesn't-matter crowd is our willingness to outsource technology. Companies are re-evaluating their sourcing strategies and have lengthened the list of potential candidates for partial and full outsourcing. Some of these include help desk support, programming and application maintenance. If we extend this trend, it's likely that we'll see a lot more hosting of even large applications – like enterprise resource planning (ERP) applications – that companies will increasingly rent (to avoid implementation and support problems). But does this trend spell the end of IT? Hardly. Outsourcing trends dovetail perfectly with commoditization trends. Companies have correctly discovered that they don't need to develop core competencies in maintaining PCs or supporting Microsoft Office – and why should they? This kind of support should be left to specialists who can offer economies of scale, reliability and cost-effectiveness.

The real story here is not commoditization, discipline or outsourcing, but the separation of technology into operational and strategic layers. Operational technology is what's becoming commoditized. Strategic technology is alive and well – and still very much a competitive differentiator. It's even possible to argue that since operational technology has been commoditized, we're finally ready to strategically leverage technology.

Operational technology enables current and emerging business models and processes in well-defined, predictable ways. Hardware price/performance ratios are perhaps the most obvious example of this trend, but there are others as well, including what we're willing to pay for programming services (here and abroad). We now expect companies to excel in the acquisition, deployment and management of operational technology. We expect them to know what they're doing here – recognizing that mistakes can be extremely costly and even threaten a company's position in the marketplace. Far from being irrelevant, given the size of our technology budgets and our growing dependency on technology, it's essential that we get the operational layer right. Many companies are very good at it. Some companies are horrible. There's huge opportunity – and risk – here. Try telling a CEO that a botched $100M ERP project doesn't matter.

Strategic technology on the other hand is the result of creative business technology thinking where, for example, a Wal-Mart streamlines its supply chain, a Starbucks offers wireless access to the Web to its retail customers (to keep them inside their stores spending money), and a Vanguard leverages its Web site to dramatically improve customer service. There's no limit to how strategic the business technology relationship can be. Again, the exploitation of strategic technology – like customer relationship management (CRM) and its personalization and customization cousins – is dependent upon solid operational technology. The same is true of wireless communications, automation and dynamic pricing. You mean to tell me that the ability to change wholesale and retail prices in real-time is not strategic?

Strategic technology is liberated by operational technology. How much time did we spend putting out operational brush fires in the 1980s and 90s? Was there any time left to think strategically? A lot of basic hardware and software just didn't work that well back then, but now technology's reliable and cheap – and now there's finally time to strategically leverage technology – so long as the distinction between operational and strategic technology is well understood. Marching orders? Solid operational technology that enables creative strategic technology. If we get this relationship right, technology can contribute directly to efficiency, profitability and growth. Thinking about business and technology holistically will help.

Organization of the Conversations

Chapter I talks about the stakeholders in the business technology sweepstakes. While we might sometimes think that finance, marketing, sales and product development professionals are disconnected from technology investment decisions, Chapter I will introduce them all as part of a management team that has vested interests in business technology decisions, since none of them can really do their jobs without business **and** technology.

Chapter II sets the stage for the business collaboration and technology integration discussions that will follow. Chapter II explains where the business technology relationship has been and the factors that are influencing it today.

Chapter III describes the collaborative business models that are appearing as fast as we can leverage technology to support them. All of these models assume connectivity among customers, employees, suppliers and partners. They include models that link suppliers, distributors and sellers (supply chain planning and management), models that use prior information about customers to personalize and customize connectivity, and models that even assume the value of completely automating transactions. Chapter III also talks about the trust that must exist for collaborative business models to work.

Chapter IV talks about the different kinds of technologies out there that enable and extend collaborative business models. Some of these include infrastructure technologies like database management, transaction applications like enterprise resource planning (ERP) and customer relationship management (CRM) systems, and communications technologies like wireless cell phones, personal digital assistants (PDAs) and virtual private networks (VPNs). The trick is to understand what these technologies do – not how they're made. We'll focus on how these and other technologies enable and extend emerging collaborative business models at a level that makes sense for managers.

Chapter V describes some alternative organizational structures you might consider as you try to make the business technology relationship productive. The chapter challenges some of the more traditional reporting relationships and even questions the need for an official Chief Information Officer (CIO).

Chapter VI focuses on business technology discipline – the processes by which we develop collaborative business strategies and buy and deploy computer hardware and software. Chapter VI focuses on business scenario development, the development of business cases, project management and return-on-investment (ROI) calculations to measure business technology investments' impact.

Chapter VII talks about the kind of people necessary to make the relationship between business and technology productive. It offers some suggestions for assessing people and measuring the skills gap between what you need and what you have. It also offers some rules of thumb for considering how much activity to outsource.

Chapter VIII wraps it all up with a picture – exactly like the one that appears below – that illustrates the synergism between emerging collaborative business models, technology that's integrating, and business technology management best practices.

A Little About the Tone of the Book

This book is written a little differently than most books about business technology. I've tried to write a book that was relatively easy to read. In fact, I tried to write a book that was almost fun to explore. The tone is deliberately

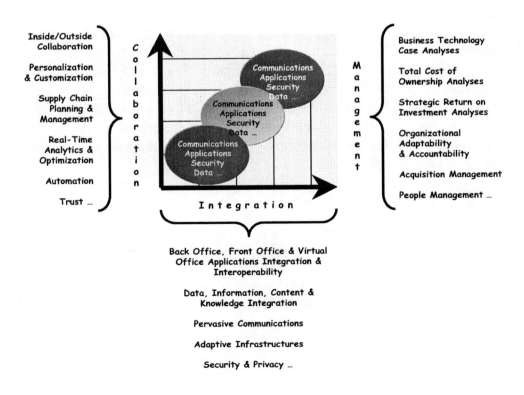

conversational. There's also a story embedded in the book, a drama that takes place in just about every corporation out there. See if you can predict the ending.

The Take-Aways from the Conversations

When you finish the book hopefully you'll spend less on – and get a whole lot more from – the business technology investments you make. I guess if the book were soap, I'd offer a guarantee that if you didn't come out cleaner I'd refund your money (but since this is the publishing business, logistics prevent any such promises). I will, however, be really disappointed if after these conversations you still think the same way you did about business technology – and the management best practices around acquisition and deployment – before you read the book. Your guarantee is my fear of embarrassment.

But when all's said and done, **The 2nd Digital Revolution** is about redefining the relationship between business and technology in your companies. While debates rage on about the virtues of outsourcing, whether or not Linux will really threaten UNIX and Windows, and if IBM can really deliver "on demand," the real challenge is to get business and technology to integrate into a whole much greater than the sum of their parts. The late Bob DiStephano – the highly respected CIO of the Vanguard Group – used to say, "There are no technology decisions – only business decisions." Bob was ahead of his time. But now Bob's time has caught up to us all. The conversations in this book should help you think about technology as business and the business of technology much more holistically.

Acknowledgments

Lots of people contributed to this book. Most of them have no idea they've done so. Andy Sage, Dean Emeritus and quintessential professor in the School of Information Technology & Engineering at George Mason University in Fairfax, Virginia really helped with the conceptual coat rack on which I'm still hanging ideas. His superb ability to describe and apply the discipline of systems engineering is still the compass I use to organize all this stuff. Bob Zito, now retired from CIGNA and United Health, helped me parse the realities of business technology management into realistic approaches. Bob is a pragmatic technologist who operates within the parameters of the environment in which we work. Much more recently, the graduate and undergraduate Capstone students at Villanova University helped me communicate the ideas in the book. I've already held the conversations in this book with them, conversations to which they've contributed significantly even as they worked me for grades. The Villanova Executive MBA students – I, II and III – in particular contributed to the book. I used the materials in class over a long period of time. These students were a tough audience. Their questions and challenges were excellent – and helped focus the discussions on the right things.

Others who have flavored my thinking over the years include Craig Fields, Bob Kahn, Nick Negroponte, Roger Schank, Bob Fossum, George Heilmeier, Bob Young, Dexter Fletcher, Jon Wilkenfeld, Jerry Hopple, Demetri Papademetriou, Len Adelman, Lee Ehrhart, Al Davis, Peter Freeman, Dick Fairley, Dick Lytle, Paul Weinberg, Mark Broome, John Pacy, Sam Palermo, Andrea Anania, Chris Pacitti, Larry Meador, Mike Burns, Anne Wilms, Jim Ounsworth, Rob Adams, Brian Dooner, Dick Guttendorf, Bob McParland, Bob Nydick, Tim Monahan, Steve Fugale, John Carrow, Matt Liberatore, Steve Zarrilli, Lee Yohannan, Jon Brasington, Jon Carrow, Joel Adler, Ralph Menzano, Jeff Worthington, Jeff Miller, Lucinda Duncalfe Holt, Vince Schiavone, Scott Overmyer, Charlton Monsanto, Frank Mayadas, Jerry Wind, Scott Snyder and Paul Schoemaker. They all contributed – whether they realize it or not - to aspects of this book. Max Hughes deserves – and demands – special mention. He read some early chapters and while he "liked" them, he also complained about their lack of "actionable" content. So I reworked the content to satisfy his – and I'm sure your – desire for insight you can actually use.

My lifelong partner, Denise, supported the entire journey. Without her support projects like this would simply never happen. I've been at this for a long time now and she's always been there. My daughters, Katherine and Emily, provided some serious grounding along the way. In their own unique ways, they influenced the form and content of this book. I wonder what they thought about dad sitting in front of a machine at all hours of the day and night spewing out word after word. I should ask them.

I'd also like to thank the Labrecque family. It is an honor for me to hold the Thomas G. Labrecque chair of business at Villanova University. Mr. Labrecque was a complete business professional and a man who's personal values and professional business decisions worked side by side for decades. Unlike many who have gained infamy over the past few years, it is comforting to know that some incredibly successful people were also some of the very best people. Mr. Labrecque was one of the incredibly successful good guys. The chair that he endowed enabled me to pursue much of the research that led to this book. Thank you, sir.

Thanks to everyone, we have a book here. Hopefully, it will help us get from one place to another. At the very least, if you read this book you'll save a ton of cash. At best, you'll position your business to succeed in the 21st century.

Steve Andriole
Bryn Mawr, Pennsylvania

Endnotes

[1] The most recent suspicions appeared in Nicholas Carr's. "IT Doesn't Matter," *Harvard Business Review*, May 2003.

[2] Nicholas Carr, "IT Doesn't Matter," *Harvard Business Review*, May 2003.

Chapter I

Who's Here?

All conversations have participants. In this case, they include all of the decision makers that directly or indirectly influence the nature and direction of the business technology relationship. It's important to understand who they are, what they believe, how they prioritize projects and the politics implicit in their roles.

It's hard to say if this is a dinner party, a corporate retreat or just one of those long meetings that happen every day. Regardless of the venue, we can describe the guest list. There's no question that some of the guests will talk more than others, and some will leave these conversations with a different perspective on who they are, what they do, and how business and technology need to converge.

So who are the participants?

- **Chief Executive Officers, Presidents, and Chief Operating Officers** who often don't know all that much about technology but who are now looking closely at rising business technology budgets and how computing and communications technology can be both tactical and really, truly strategic.

- **Chief Financial Officers** and their henchman who more often than not see technology as a giant sinkhole into which they continuously pour money, but who also want to exercise best acquisition practices as they try to reinvent themselves as offensive – rather than defensive – players.
- "Technologists" – **Chief Information and Chief Technology Officers** (and their technology architects) who buy and deploy all the gear, and how their roles are changing as they get closer to business processes and transactions – or else.
- **Marketing and other hype-sters** that are here to find nuggets they can use to brand and re-brand their companies (or steal some technology dollars to rethink the marketing strategy).
- **Chief Security and Privacy Officers** who have finally been awarded an official role in these conversations.
- Advisors that come in many flavors, such as **General Counsels and Corporate Directors**, who participate in business technology decisions and help companies keep their eye on the ball or, in the case of many directors and advisors, try to learn as much as they can – as fast as they can - about the technology platforms on which their companies sit.
- All of the **other "chiefs"** who have vested interests in what their companies are doing with business technology and who often spend way too much time building their fiefdoms instead of optimizing business technology.
- **Worker Bees** who make business technology happen, the sometimes cynical but often talented professionals who've been hardened by too many expensive, ineffective years on the job.
- **Wall Street Analysts** who cover the business models of public companies as well as the role that technology plays in the execution of those models.
- The **other "constituents"** in the business technology world, such as customers, suppliers, partners and shareholders who use the technology we deploy.

Do all of these people really have to be here? Yes, they do. The only way we'll achieve business technology convergence is if the communication channels in your company are wide and rich. You get to enjoy these conversations. One last thing: those faces pressed up against the glass looking in? They belong to the vendors and consultants. You decide if you want them inside or outside of the hen house.

Let's introduce the cast of characters.

Big Guns

There's the Chief Executive Officer, the President and Chief Operating Officer – who might be the same person (or you) – who really don't know all that much about specific technologies, though they're starting to get a sense of just how complex the care and feeding of their technology infrastructure actually is. CEOs of technology companies of course are knee-deep in the stuff, but they're not directly part of this conversation: they get to listen, but this talk is about how vertical companies – insurance, financial services, manufacturing, retail, and pharmaceutical companies – can integrate computing and communications technology into their existing and new business models and processes, not about how to help technology companies sell more gear. Vertical CEOs, Presidents and COOs may or may not have a serious relationship with a CIO or CTO. That's because in some companies the CIO reports to the CFO (crazy) or some other executive, or because many (especially smaller) companies don't even have CIOs or CTOs. Many CEOs, Presidents and Chief Operating Officers climbed the corporate ladder through sales, marketing or service, not from the technology trenches. Along the way they certainly encountered technologists and technology managers, but many of these people were way too geeky or weird for them: there's very little bonding among finance, marketing and technology in the Fortune 500. Their image of technology is not necessarily positive and almost certainly not cool. At the same time, every one of their transactions is dependent upon technology. They understand this, but they're not all that happy about it.

So they've come to this conversation under a little duress, though a voice deep inside keeps warning them about the technology budget that just keeps on growing.

There's also some unconventional pressure brewing. CEOs and senior management teams are growing increasingly concerned about their directors and advisors and their growing awareness of the both the expense and potential of technology. Sure, while most board members are buddies (and advisors are

Copyright © 2005, Idea Group Inc. Copying or distributing in print or electronic forms without written permission of Idea Group Inc. is prohibited.

wannabe buddies and wannabe directors), they're starting to ask tough questions about the enterprise resource planning (ERP) project that overran by $50M, the 5th false start to deploy a customer relationship management (CRM) application, the network and systems management framework that never got deployed, and the now defunct e-business initiatives. Years ago, failed projects could be swept under the rug or eaten as part of the larger infrastructure budget, but hundreds of millions of dollars are tough to hide, especially these days.

Finally, the big guns are here because many technology initiatives are now considered "strategic" (in spite of what Carr and others might say). Public companies are now prone to describe their big technology initiatives as strategically significant projects that can increase market share, efficiency and profitability, especially when they often know that the return-on-investment (ROI) on huge technology projects is uncertain at best. Even more dangerous is the profiling of redundant technology organizations, applications and infrastructures when companies merge. If we've heard it once, we've heard it a thousand times: "this merger will trigger cost reductions in the hundreds of millions of dollars simply due to the elimination of redundant technology." People actually used to believe these claims even though there's absolutely no evidence that any cost savings through technology elimination and integration are a slam dunk. In fact, given that most mergers fail to increase shareholder value, it should come as no surprise that it's tough to reduce technology expenses through mergers and acquisitions.

So the big guns are curious, skeptical and motivated to improve the business technology relationship. They need this stuff – just about every transaction in their company is dependent upon computing and communications technology. They're also naturally evolutionary, seldom revolutionary. Which means that CEOs – perhaps you – are inclined to under-estimate the power of a re-engineered business technology relationship. Probably the best way to get to cynical CEOs, Presidents and COOs is through cost-savings, since even technologically-challenged CEOs have heard about the big failed technology projects (though it's unlikely that they heard the whole truth and nothing but the truth about exactly why the project failed). They also understand supply chain management and Web-based self-service, among other technology-driven efficiencies. We'll hit these kinds of capabilities hard in the conversations that follow, but – like the CFOs participating in the conversations – we'll also turn these people from good solid defensive players to creative offensive ones.

Bean Counters

There's the Chief Financial Officer – who almost always shows-up at the meeting on time and well-pressed. CFO power rises and falls with the capital markets and your company's specific situation. During bull markets when everyone's making money CFOs are expected to be creative financial engineers and make the numbers look great, but when times get tough they're expected to be task-masters, the bearers of bad news. There are those who believe that CFOs are happiest in bear markets when the wheels come off their companies, but that's probably just an urban legend. CFOs know when to resurrect their calculators. When e-business was a protected strategic initiative they retreated from conventional return-on-investment and total-cost-of-ownership calculations, but since 2000 they've been out in force inspecting business plans, business cases and all of the ROI and TCO data they can get their hands on. Regardless of their motives or personalities, CFOs turn out to be essential business technology convergence conversationalists. Why? Because their undying commitment to logic makes them prime candidates for membership on the new business technology convergence team, though, perhaps because of past sins, their membership is by no means guaranteed. The questions they ask – in bull or bear times – are fundamentally the right questions: investments in technology are about business profitability. Finally, the CFOs that participate in the following conversations also have very long memories: these are the people we'll have to convince about the prudence of yet another enterprise application project (after the last several expensive failures).

But CFOs want more, they want more than a seat at the big table; they want a **big** seat at the big table. Which means they have to become strategic sometimes and not just tactical and operational, which is their natural predisposition. Business technology convergence is actually a potential hobby horse for enlightened CFOs since the best business practices of convergence are consistent with what CFOs usually espouse anyway. Does this mean that it's becoming fashionable to actually do due diligence? To actually develop sound business cases before making monster technology investments? If we're not careful these people could rule the world.

Techies (AKA Propeller Heads)

Of course there's a Chief Information Officer and Chief Technology Officer – especially because some companies have one or the other (and some have both). Since no one's completely sure about how the two work together, we'll invite them both to the conversations. CIOs and CTOs – especially if they began their careers in the 1960s – see the world schizophrenically. On the one hand, because they grew up in a profession that pinnacled with perfect standardization controlled from data centers (AKA technology bunkers) where business users of their technology had to beg them for more capacity, better service and even the slightest changes to their applications were kept locked deep in the bowls of the bunkers. It was widely known that if you angered the CIO in those days you'd never make your revenue numbers. These fiefdoms collapsed when personal computers (PCs) and then servers flew under their radar screens and began to reproduce uncontrollably. While many of these veteran CIOs have recovered emotionally, there are still lots of them hanging around shell-shocked not only from 1st generation client-server computing but from the ultimate distributed environment – the Internet and its World Wide Web.

The other side of their psyches is focused on the possibility – for the first time in their professional careers – of reliability, interoperability, modifiability and even security. These CIOs and CTOs are excited about the future – regardless of their age. We need to bring them all to the conversations since for the time being at least they're the ones that make all this stuff work. Time being? We'll have that conversation later.

Some of them are therefore salvageable – but many are not. We used to think that the best CIOs came up through the technology ranks, not the business units. The assumption was that it was easier to teach technologists about business than the other way around. Back then we were right, but now that assumption is wrong. It's better to take a successful business manager and inject him or her with technology than hope a life-long technologist embraces simple and complex business models and processes.

CTOs are another breed altogether. They often eat, drink and sleep with technology. Check the propeller head quotient ("PHQ") before paying too much attention to these people. They have to participate in business technology convergence conversations, however, because they often own your technology architectures and usually manage your technology research and development budgets. (Since I've held the CTO title I have to offer that not all of them are techno geeks, that some of them are very balanced.)

About that relationship between CIOs and CTOs: in some companies the CIO is really the chief infrastructure officer, the one responsible for making the trains run on time. When this is the case, the CTO is the chief technology visionary, the one responsible for setting the business technology investment agenda. Sometimes, however, the CIO is more than an infrastructure jockey and the CTO is the blue-sky dreamer responsible for describing how the world will work in five or ten years. When this is the case, it won't be hard to keep the CIO focused during these conversations (as the CTO floats into space). As always, it's about power. Whoever controls the most business technology investment dollars gets to talk the most, but regardless of whether it's your CIO or CTO (you or someone else), power will be distributed across several chiefs.

Hype-Sters

The Chief Marketing Officers that show up will remind everyone about how instrumental they were raising our consciousness about the Internet: this was their crowning achievement in the 20th century and they're not about to let us forget it. But they're also growing increasingly sensitive to how online and off-line marketing can get to most everyone and to how pervasive marketing can deliver impressive results. Dying for opportunities to empirically prove their value, CMOs will come prepared to contribute to the conversations. They still have unfinished agendas, and they've figured out that they can hijack "killer (technology) apps" along with their ingrained ability to defend their budgets in bear and bull markets.

CMOs and their teams are interesting – if not sometimes a little boring – conversationalists. Why? Because there's value in their disconnection from the trenches, from sales and service. The good ones are more about style than substance and their stylistic mirrors can help management understand how the world is taught to see the company, to understand its products and services. Many a manager has experienced a disconnect when watching a television or digital commercial for the products and services they sell – products and services that look very different from the ones they provide every day. Hype-sters can help us try out new styles and forms so long as they don't deviate too much from real substance and content. In other words, the natural, sometimes healthy tension among sales, marketing and service can be a rich source of insight about next generation product and service development and support.

We've also learned that marketing hype – especially in the digital age – can quickly strategically position and re-position companies that need identities. It's much easier to develop and communicate brands today than it was a decade ago. Your hype-sters can help a lot, so long as they take a broad view of their roles. Yeah, they should stay.

Protection

It's about time that we acknowledge the importance of security, privacy and business recovery. For a long time, they were the poor relations of big boy budgets. I personally had trouble defending investments in security and especially budgets for business resumption planning. These are "low probability events," I was told. What a difference 9/11 has made in the way we think about security, disaster recovery, business resumption planning and even privacy, and there's a new appreciation for what the military has worried about for decades: information warfare.

So the conversations here must include the Chief Security Officer (CSO) and his or her compatriot, the Chief Privacy Officer (CPO). But these people are a little weird. They're like the prophets who predict Armageddon but don't really believe it's going to happen – and when it finally does, they're unprepared to deal with the breadth of the problems they've been describing all these years. Our understanding of what security and disaster recovery really require will grow with the budgets necessary to satisfy these requirements.

Consigliore

Let's not forget the directors of the company and their advisors, or the General Counsel. The chairs of the Board of Directors and Advisory Board may also play special roles in this conversation, so long as we can: (a) find them, and (b) wake them up. Why so harsh? If we've learned anything over the years it's that hand-picked boards are, well, hand-picked boards. Senator Lieberman's comment that the Enron Board of Directors not only fiddled while the company was burning but also roasted marshmallows in the flames, sort of says it all. Not that all boards are bad or unethical or anything else nefarious we can think of, but they are incentivized not to challenge senior management, the hand that feeds them: board compensation for service to large public companies is non-trivial. It's therefore unrealistic to assume that directors will challenge management too aggressively. There's a reason they call them "good 'ol boys." Of course if the boys all get together after hours and decide to whack the CEO, then all bets are off, but that conspiracy usually takes years to plan and execute.

This is all unfortunate given the conversations we're having here, because technology investments are strategically important and incredibly expensive, precisely the kinds of investments that boards should evaluate. How often do CIOs make board presentations on the status of major technology projects? How often do we let CTOs talk to directors? (Actually, we probably should keep CTOs away from them.)

Lots of boards – though of course not all – are pretty clueless about business technology convergence. But they must participate in these conversations. They must widen and deepen their understanding of technology capabilities and trends and how the intersection of business and technology is the great, untapped whole-that-is-greater-than-the-sum-of-its-parts. They must also participate because board accountability is expanding, as quickly as directors' and officers' (D&O) liability insurance premiums are rising. Boards need business technology dashboards to keep track of large projects and should inspect the business cases that launch mega projects. Period.

Advisory Boards are popular these days for several reasons. They represent a company's farm team from which new directors can be drafted and they also increase a company's visibility and "optics": it looks good to have heavy hitters hanging around companies. The assumption is that they'll help the company with sales, marketing or product development by making calls to their well-placed buddies. Smart companies pick their advisors according to their ability to really help the business, not according to their golf handicaps. Advisors need to participate in these conversations because they need to understand the potential of business technology convergence and because they can sometimes lead the conversation. Even though they're wannabes, they tend to be less beholding to the senior management team that brought them in and are therefore freer to speak candidly about business technology.

And let's not forget the General Counsel, ideally the real consigliore in the company. Good General Counsels have high ethical standards and are capable of standing a little outside of your company, capable of reasonably objective assessments about what's working and what's not. In addition to their legal duties, they define a company's professionalism. It's a good idea to keep them busy on positive things, instead of messes that need cleaning up. Think about all of the calories that got burned at Worldcom, Enron and Global Crossing, among other companies, calories that could have been targeted at new markets and improving profitability. I tell my students that ethical behavior is "right" **and** good business: you only get to lie a few times before you get tagged as a liar – and liars usually can only do business with other liars.

All in all, consigliore need to participate in the conversations that follow because they represent intellectual assets that can be leveraged in all sorts of ways to help your business. They need to understand and support your commitment to business technology convergence. In some cases, they can lead the charge. And so long as we deal with their incentives and motivations, we can screen good advice from self-serving tripe.

Other Chiefs

Have you noticed how many chiefs there are out there? In addition to all of the well-established chiefs, we have Chief Learning Officers (CLOs), Chief Knowledge Officers (CKOs), Chief Partner Officers (CPOs) and Chief Alliance Officers (CAOs), among any number of others. What the hell do all of these people do? And why do they all have to be Chiefs? If you take a hard look at your company and its chiefs you can quickly determine if they're misnamed by applying the Chief Test, which simply measures the power/responsibility ratio. If the Chiefs have lots of responsibility but no power, they're bogus Chiefs, but if they have both power and responsibility, they're legitimate (though you still have to figure out how to manage all of the tribes they control). What's going on at your company? Do you have lots of chiefs? And do they have lots of responsibility but no authority? Are you an empowered chief? Or do you live each day in fear that a major project will be thrown your way over which you have very little control?

Legitimate chiefs need to be part of the business technology convergence conversation. Bogus ones can sit in, but really shouldn't say too much.

Worker Bees

There are all sorts of worker bees flying around your company. These are too often the faceless, nameless managers and their staffs that make it all happen. Directors are often worker bees, though they're usually a little higher in the chain of command. They all need to at least hear these conversations. The smart ones should be encouraged to speak up, really encouraged, since they hold the execution keys in your company.

How often do we forget this? I remember countless new initiatives kicked off by senior executives who flew in to give uninspiring pep talks to the troops, who had heard it all before. I also remember initiatives that were launched and sustained by executives who sincerely believed in what they were asking the troops to do. Needless to say, the latter often survived – the former almost always crashed and burned.

Without the troops, business technology convergence won't happen. Hopefully, the conversations that follow will persuade them not to game the initiative but to commit to it. Let's make sure we include them in all of the discussions.

Constituents

How about the customers, suppliers, employees, partners, vendors, shareholders and venture capitalists that round out the guest list? While they might not be front-and-center during the conversations we're about to have, their issues and concerns must be represented.

At the top of the list are the customers that buy all the stuff we make or consume the services we provide. The business technology that you deploy directly affects not only your profitability but your customer service as well. Your ability to compete is also driven by the effectiveness of your business technology investments. Business technology touches every customer directly or indirectly.

Suppliers have a vested interest in how all this goes. Everyone benefits from an efficient supply chain and the right tools can make supply chain planning and management work for everyone's benefit (including the software vendors that supply the code).

Employees have lots to gain as well. When business technology is well-deployed it means that portals, Intranets and remote access to your networks and applications become part of your connectivity repertoire. You can save money by deploying a lot of this stuff (though of lot of it is really stupid). Well-trained employees can optimize your business technology investments.

Everyone has partners. Not just in your supply chain but across all of your manufacturing, distribution and service channels. These partners can participate in product development, marketing, co-branding and customer service. Here's an example of how weird partnerships can work. Dell sells machines online and they bundle Microsoft software and HP printers with their configurations – or at least they used to. They need this stuff to make the package work. But they also compete with HP in the desktop and laptop computer markets. Managing these channels for mutual benefit is tricky – and getting trickier, especially when you calculate the margins within the value chain of the partnerships. It's possible, for example, for one company to make much higher margins on someone's products than their own. If you make more money channeling a competitor's product, how should you manage that partnership? Sometimes it breaks down. The Dell/HP printer partnership blew up in 2002.

The technology vendors, as well as the consultants that sell new business strategies and deploy technology, have a whole lot to gain – or lose – from the conversations in this book. Look, they make a lot of money. The United States spends a trillion dollars a year on hardware, software and services. Some of the vendors that supply strategic consulting bill themselves out at over $300 per hour. Our friends in Redmond have billions in the bank, no debt, and an annual research and development budget that exceeds $5B. IBM spends over $7 billion on R&D and has their finger in all of the technology pies: arguably, they are the first and only "total solutions provider," since they can provide hardware, software and services organized impressively around vertical areas of expertise to medium and large businesses. SAP, Siebel, Oracle and other enterprise software vendors partner with consultants such as Accenture, IBM/PriceWaterHouseCoopers and KPMG to implement and support their applications. Cisco needs to sell more routers and Palm needs to sell more personal digital assistants (PDAs). They all have constituents, too. But business technology convergence requires us to think about technology and strategy vendors very differently than we have for the past few decades. They have a decision to make: they can become full business technology partners or they can continue to pursue their covert and overt adversarial sales and marketing practices that have them on at least half of the wanted posters in corporate

America. They need to listen closely to the conversations in this book. They can still make tons of cash but the way it will happen must change.

Shareholder value. What a concept. Given what's happening out there with class action lawsuits filed by angry shareholders against companies and the analysts that cover their stocks, maybe it's time to rethink what the concept really means. Let's face it. There's lots of subtle and not-so-subtle manipulation of the capital markets. Financial reporting is often, shall we say, "incomplete." Are shareholders constituents here? You bet. Business technology convergence will help increase the value of the companies in which we invest, in which we believe.

Theoretically, the Wall Street analysts that cover public company stocks represent shareholders – and their vested interests. While technology has not played a huge role in the development of corporate valuation models, it will begin to contribute directly to what analysts think are one of the major drivers of corporate wealth.

Last, but not least, are the venture capitalists that fund some of the companies with vested interests in business technology convergence. They fund all kinds of companies including vertical and technology companies, as well as consulting companies. They are simple, one-celled organisms: if a company smells like a return on an investment, they'll write a check. They are mutant constituents since they usually have short-term objectives and really don't care all that much about the impact their companies' products and services have on the marketplace, so long as they make money that they can distribute to their limited partners. Like the technology and strategy vendors, they only get to listen to the conversations here. While it's well beyond the purpose of this book to change venture capitalists, it might be useful for them to listen to how business technology convergence will affect the acquisition, deployment and support of business technology, if for no other reason than to help them invest in the right business technology models.

Herding Cats?

The Super Bowl commercial a couple of years ago showing a bunch of cowboys herding cats is not unrelated to these conversations about business technology convergence. Some of the guests are, of course, hopeless. No matter how profound or moving the conversations, they'll remain unenlightened. Others will only interpret what they hear in terms of personal gain, never fully appreciating the obvious logic of getting a smaller piece of a much bigger pie. But others will find the litter just in time to congregate with fellow cats long enough to see the light. Yeah, yeah, it's a weird metaphor, but it may just be possible to herd a few cats once in a while.

Chapter II

The Convergence Conversation - We're Finally Ready

If you don't want to read this whole chapter, here's the essence of the following conversation. But if I'm going to practice what I preach, I have to give you an overview of the conversation we're about to have. I guess this is what we'd call an Executive Conversation Summary. While I'd still love you to participate in the conversations, here are the main points:

- **We misinterpreted capital spending in technology in the 1990s and 2000.** We inferred that the spike in spending launched the "digital revolution." In fact, it was the result of Year 2000 fears, e-business frenzy, and investment subsidies provided by start-up companies fueled by private equity venture funding.
- **We've been led down the "business technology alignment" road** – a road I actually traveled – by analysts and practitioners that see the business technology relationship as sequential, not holistic – which is not how the relationship should work.
- Just about everyone I know **way over-reacted to the bursting of the dot.com bubble**. We threw countless beautiful babies out with the bathwater.
- For 30 years, we wrestled with **immature computing and communications technology** and with unprofessional acquisition and management

- practices. Did hardware and software vendors – and their partners-in-crime, consultants – really earn all the money they made in the 1990s?
- The irony is that **computing and communications technology is really starting to work.** The stuff is more reliable, more stable and more capable than it's ever been and it's now time to begin to think strategically and not just tactically about the role that technology can play.
- It's **time to think about the 2nd digital revolution**, how business will collaborate and how technology will integrate building on a business technology infrastructure that is surprisingly solid.
- **The big – sane – questions should direct this revolution.** Crazy, goofy, chaotic directives were part of the 1st digital revolution which got us – kicking and screaming - to where we are today, but should play no role in our exploitation of maturing business models and computing and communications technology.
- Bottom line: **business technology convergence is more than possible if we can wake up the right adults**.

Let's take a closer look.

What Happened?

This book is about business and technology. It's not about "business technology alignment," or technology best practices or how to clean up your **existing** technology organization. The conversations here will not excite rabid technologists, though many technology managers will be unable to control their enthusiasm (if only because they're included). Nor will the conversations warm the hearts of too many consultants or technology vendors, though there's a potential partnership embedded in hundreds of creases throughout the conversations about the convergence of business and technology. If you're a C-level executive, a member of a corporate Board of Directors or Advisory Board, or a line of business or technology manager, you might find a lot of the conversations just what you need at this flash point in time. If you're a consultant or technology vendor, you might want to shadow the conversations very closely. If you can find a Groucho disguise, this is one party you want to crash.

Let's begin with some candor. While it was possible just ten years ago to think about business without an immediate reference to computing and communications technology, it's no longer sane to separate the two organizationally or functionally. In fact, it's impossible to conceive and deliver new business models without defining the role that technology will play, a role that in many cases will be the dominant role. The implications here are troubling – especially because of the people in your organization to whom you've given responsibility to think strategically, and because of our historic approach to technology management – which has been biased to the back office for decades. How many of the professionals in your company understand the range of computing and communications technologies at your disposal? Does your technology organization sit at the head table in your company, or do you only call them when the networks crash? The history of technology organizations is inconsistent with the business technology convergence that's occurring, a trend that's actually dangerous to organizations that don't break completely from the inertia that created data centers, help desks and server farms. Business technology "alignment" – if we should even still use the term – is no longer about trying to get technology consistent with business strategy and tactics, but about the seamless intersection of business, technology and management.

More candor. In spite of what the business technology literature claims, there have not been a lot of revolutionary or "disruptive" technology events over the past 30 years. The list – from a C-level perspective – includes relational database management, e-mail ubiquity, the end of the mainframe era and the beginning of 1st generation client-server or "distributed" computing, the Year 2000 compliance challenge, and of course the Internet's morphing into the World Wide Web. Consultants and vendors love to talk about the next new thing because they have financial vested interests in your always wanting to buy new stuff, migrate to new hardware and software and especially undertaking huge enterprise-wide initiatives. Sometimes these initiatives make sense and sometimes they're hideous. Depending on your perspective – and age – you'll take – or pass on – the bait. The key is to conduct due diligence that assumes a full partnership between business and technology, making certain that the divide-and-conquer, separation of powers approach to technology investing that so many of us have practiced over the years abruptly ends.

Lots of us believed that the Y2K problem was a non-problem, that computers and computer-supported devices would not fail on January 1, 2000. Lots of you probably had the same feeling. But the hype was maddening. What worried us most, however, was the incredible spike in technology spending driven by the need to make hardware and software Y2K compliant, and how lots of us misinterpreted those investments as an unambiguous signal that information technology (IT) was driving all of the new business models and processes, especially e-business. Well, guess what? That's not what was happening. Savvy technology managers saw the deployment of enterprise resource planning (ERP) applications, the migration to compliant operating systems and the replacement of older hardware platforms as a cost-effective alternative to rewriting and rewiring old gear – and they were right. While it might have been cheaper in the short run to just redo some COBOL code, it made no sense in the long run since the newer hardware and software solved the Y2K problem **and** offered tons more capability (assuming, of course, that you could get it to work). The money spent during the 1997-2000 was largely to achieve compliance, not to buy tickets to the new digital revolution.

Around the same time, e-business spending joined Y2K compliance as a "protected" budget line. You simply had to have an e-business strategy – even if you really didn't know what you were going to do with it. This second spending spike also confused everyone, as e-business and Y2K spending contributed to the greatest capital investment wave that computing and communications technology has ever seen. Taken together, how could we have interpreted the investments as anything but a revolution? We were wrong, and lots of people knew it, but CNBC, the big (and boutique) investment banks, and especially private equity hype-sters proclaimed the arrival of the unstoppable new, digital economy. After a while, it all seemed to make sense. Of course, in retrospect, it was silly.

Oh yeah, there's one more thing. The period from 1995-2000 provided free capital to technology start-ups. Venture capitalist flooded the market with cash raised from general and limited partners eager to cash-in on the dot.com phenomenon. Young companies hit the market with huge VC subsidies that underwrote incentives for "real" companies to try new computing and communications technologies. Everything was half-price, so adoption rates actually looked healthy – until no one could find any returns on the investments and the subsidies dried up.

It was the perfect storm. The probability of these three trends colliding was miniscule. But collide they did, and lots of us made inferences about the future that would turn out to be flat wrong.

Figure 1 paints the picture.

Let's stay candid. While lots of technology worked well during the 1980s and 1990s, lots of it didn't. Computing and communications technology has only

Figure 1. The perfect storm

started to "work" during the past five years or so – so your instincts about cost-effectiveness and return-on-investment were sound. At the same time, "working" means that the industry has finally begun to think seriously about the integration and interoperability of different hardware, software and communications systems, and that reliability is now approaching levels similar to that of home appliances, automobiles and air conditioners. The auto example is interesting because of that industry's own struggle with quality. It wasn't too many years ago that many Americans felt so strongly about the American industry's inability to deliver or support quality products that they turned aggressively to Japanese, German and Scandinavian auto manufacturers for better cars. The American automobile industry had no choice but to respond to the quality challenge (though some believe they never recovered from the early price/quality wars). The technology industry delivered reasonable quality when their systems were simple, when mainframes ruled the data centers and the number of hardware and software manufacturers was relatively small. But as the industry grew, variation grew to the point where hardware and software incompatibilities were more the rule than the exception, forcing technology managers to spend incredible amounts of time making all the pieces work together. Like the American auto industry, the technology industry has finally started to listen.

This conversation is occurring at exactly the right time. Many of us have struggled with the relationship between technology and business for decades. Why does technology cost so much? Why are we always buying new gear?

Why doesn't this stuff work as advertised? Do we really need a CIO **and** a CTO? What the hell is a CTO, anyway?

We're at a flashpoint. The pace of technology deployment and business velocity has already outstripped our ability to assess its impact on how we live, produce and distribute. It's now time to think about where all this is going – and how to optimize it. Let's acknowledge the following:

- We've lived through several computing and communications "waves," all the way from mainframe-based computing to business processes that only exist on the Web delivered by distributed servers.
- Lots of the early stuff did not work as advertised, resulting in a cynicism about "business/technology alignment."
- The bursting of the Internet bubble resulted in an over-correction, a dubiousness about all varieties of distributed computing effectiveness.
- Just when the stuff started to work and just when business models were beginning to morph out of their traditional vertical silos, capital spending in technology collapsed.
- In spite of all these reactions and trends, we're now sitting at one of the most important crossroads in the history of business technology, and especially business technology management.

Let's assert that the computing and communications technology we've developed and deployed over the past 30 years represents a kind of prototype. PCs have gotten cheaper. Companies have access to the Web, and we're now free to think about customer relationship management, Web services and intelligent agents. But we still struggle with a lack of standards, integration and interoperability problems and chronic disconnects among our back office, front office and Internet applications. We've gotten good at creating technology pieces, but we're only now beginning to focus on how they all work together.

Let's also assert that business models are evolving, morphing and accelerating faster than ever before. Your company now finds it hard to draw old lines around what it does and how it operates, or around what we used to call it's core competency. In fact, the whole notion of core competency is now necessarily confused since companies seriously began rethinking their supply chains, partnerships and alliances. Here's a thought: is Dell a computer manufacturing company or a supply chain planning and management company?

How hard is it to imagine Dell as a company with multiple lines of business including one that sells supply chain management software and services? The other lines of business might focus on manufacturing and distribution using Dell's advanced software platforms. Dell's core competency in this scenario? Supply chain integration.

The last 30 years constitute the first digital revolution. The next 30 will define the second one.

Companies that treat the interplay between business and technology as a simple extrapolation from even the most recent past (yes, that includes our initial infatuation with the Web) will be out-maneuvered by companies that see it all as a revolution enabled by pure business technology convergence.

The convergence of technology and business change is what's different. But unlike a perfect storm when lots of weird things happen at the same time, what we now have is a perfect opportunity fueled by the convergence of technology that's finally ready for prime time, business models that embrace speed and flexibility, and management possibilities that will treat the relationship holistically. Stated a little differently, it now almost works as advertised. Within three to five years, the truth-in-advertising gap will close (almost) completely.

Of course, you might argue that "almost" is not good enough, or that you've heard it all before, and that the most prudent approach is to simply wait until the stuff starts really working – as advertised. You can take this approach but the problem is that when convergence hits full stride you'll have to scramble to catch up to those who saw it coming. Remember the Internet: have we forgotten that Microsoft actually missed it, only to respond with the now famous "extend and embrace" initiative? Need some other examples? Remember the Digital Equipment Corporation (DEC) and Ken Olsen's famous "why would anyone want a computer on their desk" comment? Or Encyclopedia Britannica's initial rejection of CD-ROMs?

If all this is true, then the way we approach business and technology modeling should change dramatically. The approaches we've taken to business/technology "alignment" served us well for a while, but grossly miss the point of holistic modeling. (If you have any conventional "alignment" initiatives in progress, kill them.)

So how should we proceed? If a picture is worth a thousand words, here's one that hopefully communicates the essence of what I'm talking about and what this book describes.

Figure 2. How to think about convergence

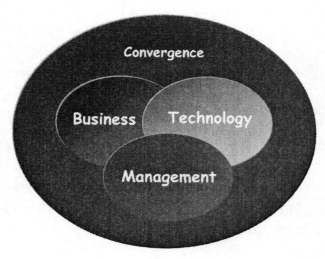

What does this mean? First, it means that organizational distinctions between business and technology should disappear and be replaced by seamless interconnections that make it impossible to address one without the others. It also means that chief generalists (CEOs, CFOs, CIOs, and even CTOs) need to become wider **and** deeper, redefining the whole notion of generalist practitioners. Can you really lead a company if you don't understand technology architectures and applications? Can you enable a company if **all** you understand are architectures and applications?

The relationship between technology and business has evolved in a way that's served each community only pretty well over the past 30 years. Companies spend millions and in some cases billions of dollars a year feeding each community that essentially co-exists for the greater good – profit, bonuses and shareholder value. By and large, the relationship "works," though there's increasing evidence of dysfunctionality in the trenches where wars continue to erupt among technologists, business managers, and finance professionals, the latter group for the life of them cannot understand how an enterprise resource planning (ERP) implementation project in a Fortune 500 company can cost a $100M and take three years to complete, or why in the same company it takes two years to migrate from one desktop/laptop operating system to another.

These wars aside, we now have lots of applications, databases, devices, communications, security – and even the means to resume business if a disaster occurs. Perhaps amazingly, this stuff works reasonably well most of the time.

But analysts like Paul Strassmann tell us that we're overspending on technology and the relationship between technology investments and productivity is anything but clear.[1] Others – especially in bearish economies – resurrect total cost of ownership (TCO) and return-on-investment (ROI) calculi to derail big technology initiatives. Bottom line: the relationship is still pretty adolescent.

But things are improving all the time. We whipped the Year 2000 problem, connected just about everyone to the Internet, and have begun to more deeply appreciate the need for integration and interoperability even as proprietary vendors make it tricky to make all of the pieces fit together. So life in the trenches is – well – pretty good, and the potential is skyrocketing.

Let's continue with the assertion: we've reached the point where if we continue along the same business technology relationship path, we'll undermine the very business models and processes we're trying to define and deploy. Worse, the current relationship will eventually collapse under its own weight due to organizational ambiguities, technology complexity, and our inability to satisfy business-to-consumer (B2C) or business-to-business (B2B) requirements that are appearing and changing faster than they can be satisfied.

While the evidence tells us that computing and communications technology have made enormous strides over the past 30 years, and we now routinely, if not obsoletely, talk about business/technology "alignment," technology optimization and how companies can extend their business models through pervasive communications. Our current discussions about supply chain planning and management, collaborative forecasting and automation all assume a business technology relationship that's fundamentally different from our notions of "alignment" or our organizational attempts to get technology to the head table. Discussions about whether CIOs and CTOs should report to the CEO or CFO are really very 20th century, since everyone now knows that CIOs and CTOs (or whoever's your point of contact for business technology) should breakfast with the big guys (assuming, of course, that they're house-broken).

From another perspective, it's no longer possible for a chief executive or any sane senior management team to conceive of a new – or even think of extending an existing – business model without addressing technology requirements, capabilities and costs. Some of these models are actually created in reverse, where business models extend from what's technologically feasible, not necessarily from solid (read: profitable) business models. Remember the dot.coms?

We've been through a lot over the past few years. We discovered the Internet, successfully managed the Y2K compliance problem, hyped Web-based business models, and confused even the most loyal technology investors about what drives capital technology spending. I'd argue that this confused picture is at least partially the work of amateurs – venture capitalists, Wall Street technology analysts, 20-something technology entrepreneurs and too-easily-swayed corporate technology buyers – who unknowingly conspired to distort events for their own special purposes (gee, I wonder what they might have been?). Real business technology pros were relatively quiet during this mayhem, but now – since the amateurs self-destructed – have a clear playing field. They can now step up their 21st century games.

A longer view – way back to the 1960s – saw the introduction of "data processing" to industries that barely knew what to make of computers, software and databases. The 1970s took us to a much higher level where mainframes got a little flexible, minicomputers arrived for the frugal, and PCs began to procreate among parents – like Sinclair, Kaypro and Osborne – long since extinct.

During the 1980s everyone absolutely had to have a PC at work and increasing numbers of us had to have them at home. Even personal software got easier to use, principally through Apple's introduction of the Macintosh, though business applications continued to only slowly evolve. Most of us were using home-grown software systems to keep the books, track inventory and manage our people.

The 1990s gave us client/server computing – our first real freedom from mainframe architectures – the Internet, the World Wide Web, multi-tier applications, data warehouses, data mining, applications integration, online exchanges, new security requirements, privacy issues, virtual private networks, application service providers, content management, knowledge management, network services, Java, Perl, Linux, customer relationship management (CRM), interactive marketing, Bluetooth, 802.11(a/b) and a whole lot more. Lots of us think we've achieved a new level of "alignment," or the process by which technology supports business. The truth is that the questions that dominate this new alignment consciousness are now the wrong questions. Why? Because technology and business are no longer even "equal" partners – they are an integrated whole. Technologies without business models are as useless as business models without technologies. This is the key point: technology, business and management can no longer be treated in relative vacuums. They work together or not at all. If you or your senior people don't get this you're in some trouble – because there are players in your market that absolutely do.

The approach we've taken to "aligning" information technology with business models and processes served us well until the year 2000. While most companies never quite got there it's time that they stop trying to win an old war that no longer matters. Not because the goal was wrong, but because it's no longer consistent with the business and technology trends that are upon us. The pace of technology and business change has forever altered the way we should think about how we find and service customers, suppliers, employees and partners, and organize ourselves to compete.

The net effect of all this is that we continue to spend far more on technology than we should, ask the wrong questions about technology, and consequently miss opportunities to leverage technology onto our business strategies, models and processes. We continue to march to agendas set by consultants and – especially – vendors. We also tend to default to conservative interpretations of what's really outside the box. Very few companies are serious about radically changing – or even challenging – their existing business models. Unfortunately, present conditions require much more than conservative extrapolations of current business models or the selective application of "new" technologies. Stop for a moment and assess the latest cocktail party buzzwords. How often have you heard about customer relationship management (CRM), Web services, and wireless communications over the past 12 months? What will the list look like next year? (And by the way, how many people at these cocktail parties actually knew what the hell they were talking about?)

While there are certainly organizations that suffer less than others, the lion's share of companies older than 20 are in serious trouble. Really scary is the number of executives unaware of just how serious their situation is. Huge numbers believe that Microsoft, Oracle or IBM have the answers to their problems, or management consultants can set them straight, or that in-house people that support the status quo can somehow solve them. The amount of waste is staggering. Many companies don't even know what they're spending on business technology, so they end up living ignorantly about what they could save, while others collect benchmarking data but avoid the tough questions about business technology optimization.

All of this occurs as the promise – **and reality** – of business technology is at an all-time high. The **producers** of technology products and services wax damn near poetic about what they have made possible - and what they plan to do next. Executives crow about how their companies are "upgrading" their "infrastructures" and deploying "state-of-the-art" communications and computing "architectures." But precious few really understand their own speeches. The consumers of technology are thus at a distinct disadvantage that's systematically exploited by producers. The **facilitators** of technology – the consultants – play both sides, offering advice to harried, perplexed consumers **and** the producers of technology, brokering the relationship often with the finesse of a magician – and sometimes with the margins of a pornographer.

Comptrollers are forever writing checks to buy more computers, more telecommunications, more software and more technology professionals. But Chief Executive Officers want to document the return on their investments. Chief Information Officers find themselves on the defensive far more often than in the winner's circle. Who wants to be at the head table if you always get served last? Is your technology expense "managed"?

The technology marketplace is one of the largest and fastest growing in the world today. But everywhere one looks, everywhere one goes, and over and over again in the technical journals and trade publications we still see references to the same issues, problems and challenges: "the software crisis," "the requirements problem," "the return on investment challenge," and "total quality software management," among all sorts of other dopey things. Many of us wrote about requirements problems two decades ago, we called for project "dashboards" for managing multiple projects, and committed ourselves to "process improvement." Well, here we are 20 years later asking the same damn questions and – worse – proposing the same damn solutions.

But these are "tactical" problems, problems that are created by - so they can be solved by - the providers of technology products, systems and services. Is this a conspiracy? You bet (though we could argue about how conscious it really is).

Tactical and operational problems are easy to identify. Strategic solutions are harder to come by. There are legitimate reasons why there are more books about problems than solutions. Perhaps the most obvious is the "moving target syndrome": as business requirements change, technology changes. As technology changes, price/performance ratios change. As price/performance ratios change, corporate cultures change. As corporate cultures change, global competition changes. As global competition changes, profitability changes. As profitability changes, the technology market changes. And so it goes.

What Do You Do First?

Enter the consultants. There are "conventional" consultants, "contrarian" consultants, and consultants who have solutions for problems yet to be invented. And there are vendors – thousands and thousands of vendors. Consultants and vendors seek to reduce problems to their simplest terms – not because problems are by nature cooperative, but because it's the only way they can appear confident enough to convince CIOs, CEOs and CFOs to spend more money.

This conversation assumes that problems and solutions cannot be traced to computers, management, software, people or networks – but to all of above and then some. We're no longer in the age of disembodied solutions to anything; we're in an era of complexity, integration, synergism – and convergence. It no longer makes any sense to hire a consultant who knows just about everything there is to know about software but very little about hardware, or, worse, to hire consultants that are technologically deep but shallow about vertical business models. Would you hire a car mechanic who doesn't drive?

The conversation also identifies principles that define processes that point to practical methods. Not long ago I received a call from the CFO of a Fortune 100 company about to write a check for $30,000,000 for a network and systems management framework. I asked if the requirements analysis was able to profile the organization's computing assets and network management needs, if the in-house technology professionals had performed trade-off analyses of several alternative frameworks, and if those who would actually be using the framework (to presumably manage their networks better, faster and cheaper than before) had ever used similar tools to help solve network management problems. The CFO asked me to explain what requirements analysis was, the CIO had no idea what network management point solutions were already in the organization, and the network operations center manager had not compared the new network management environment with anything else (but liked the vendor's brochures). No one had even talked to the network management professionals who would actually use the application. This short story illustrates how principles, processes and methods can be ignored - and how some relatively simple steps can lead to enhanced productivity and cost-effectiveness.

I am sorry to say that the conversations here pull no punches. Actually, I'm not sorry at all. Someone had to cross the line; it might as well be me.

The analyses and recommendations here are anchored in field and case studies not as irrefutable evidence or documentation, but as points on a new compass. Over the past 30 years we've cataloged problems and documented successes. But remember that 30 year-old cases are about as relevant today as a single anecdote about a guy down the block who had success with approach A, consultant B, or vendor C. The key lies in the extent to which generalizations hold against the moving target backdrop. For example, how can business technology investment decisions be made independent of business technology forecasts? The argument here is that any rational approach to business technology is multidisciplinary, anticipatory, adaptive **and** cautious. You see, this is not a conversation about "early adoption" of unproven technology. Instead, the conversation will hopefully get you to think differently, creatively yet soundly about business technology acquisition and deployment.

So what happens if you participate in these conversations? If you're a CEO you'll be armed with questions that should be asked of your business technology professionals – at the same time and in the same room. You'll also gain insight into one of the largest, most voracious – yet potentially most important – sink holes in history. A strategy – complete with tactics – will also be developed. If you are a CIO or a CTO, you'll receive some tactical and strategic insight on technology acquisition, deployment and management. You will be cautioned about repeating the mistakes of your competitors. You will think twice before authorizing big technology buys.

Regardless of your role, the conversations will provide you with a new perspective – an analytical compass. The central theme is simple: in spite of all of the hype, all of the serious technology and all of the rapidly changing business models, we're at a crossroad. It's now time to rethink the business technology relationship and move it from a less-than-equal to equal-partner model, to an integrated holistic one. The objective of the conversations is to help you construct a business technology convergence plan that will work for your organization, your people, your corporate culture, and your resources.

Business Technology Convergence

True business technology convergence assumes that all discussions about existing or new business models and processes will occur with immediate reference to technology and the best management practices necessary to integrate all of the pieces, and vice versa. With that in mind, let's look at the pieces and how they should be assembled.

If a company doesn't understand its competitive advantages and its current and future business models, it's doomed. Not only will it fail in the marketplace but it will waste tons of cash on technology along the way. We used to ask: "What's our core business?"; "What do we do well?"; "What markets do we 'own'?" The new questions are different:

- "What do we do profitably today?"
- "What are the profitable and unprofitable pieces of our value chain?"
- "What should we do tomorrow to make money?"
- "What will our collaborative team look like?"
- "How does technology define and enable profitable transactions?"
- "What business models and processes are **underserved** by technology?"
- "Which are **adequately or over-served** by technology?"
- "Which technologies can drive whole new business models?"

Let's look at business models and processes, technology (and management) holistically. Figure 3 offers some ideas and key questions.

It's all about the big questions. Do you know what you do well, poorly and with whom you compete? Have you thought about what your business will look like in three years? Have you segmented what you do according to margins? (One of the more interesting things about the HP/Compaq merger were all of the arguments over control of a low margin, commodity [PC] business.) Larger

Figure 3. Some key business, technology and management questions

questions include the long-term survival of your business via partnerships and alliances, the rate at which you can really change, and the interplay among business creativity, technology delivery and management efficiency. The new HPQ makes its serious money from selling ink cartridges and high-end servers.

If your company is like most, those who run "strategic planning" are often one step away from retirement, or worse. But convergence requires that business creativity be taken seriously every day, and that those who define and engineer innovation are also good strategic technologists and managers. If they aren't, you'll miss the convergence opportunities occurring as we speak.

Do you know if your technology infrastructure, applications and support all "work"? Do you know if they match your business models and processes? Do you know how the pieces might break? And, most importantly, do you know if your technology can grow with your business creativity – and vice versa? If someone asked you if you had too much or too little technology what would you say? Would it be easier if they asked if you had the right or the wrong technology? Or good people?

Who owns creativity? Who owns technology? You've made your first mistake if they live in silos that seldom communicate. (When we talk about organization later, you'll see just how dangerous business/technology segmentation actually is.) Who manages the integrated process? How is success and failure defined and measured?

Here's the benchmark: if you develop "new" business models (or improve existing ones) and then ask technology if it can support the changes, then you are sub-optimizing the business-technology relationship – and you're likely to over- or under-spend on business technology initiatives. Why? Because business models cannot exist without enabling technology and technology's only purpose is to support business models and processes (unless of course selling technology and technology services is your business). Yes, the implications here are huge. Without good synergism, you'll end up with too much, too little, wrong, expensive and unreliable technology supporting business models that may or may not exceed their potential.

If we've learned anything over the past few years, it's the importance of pervasive, secure, reliable communications. It not just about the Internet. It's about communications inside and outside of your firewalls and it's about mobile communications. It's about communications among your employees, suppliers and customers – and even your competitors. Have you ever wondered about Dell's (and other) online computer sites that sell Microsoft software and HPQ printers? (Though Dell's relationship with HP is more strained than it was prior to the HP/Compaq merger and Dell will not continue to directly channel HP printers.) Since all of these vendors need each other, they need to communicate.

It's no exaggeration to say that communications technology will make or break your ability to compete. There are all sorts of issues, problems and challenges that face your organization as it wrestles with its business strategy, its communications response and it's ability to adapt quickly to unpredictable events.

Figure 4 identifies some of these synergistic issues and questions.

Assessments need to be made first about who will connect to your "network." If your network will be wide – lots of employees, customers, employers, partners – then you may need to completely re-architect your communications infrastructure. You should simultaneously ask questions about the applications (like e-mail and workflow) that ride on the infrastructure, and how you'll manage infrastructure migration and measure communications efficiency. Yes, these are C-level questions now.

Note again that the distinction between business and e-business is gone: all business technology plans should assume full connectivity among a constantly growing number of participants. If you fast-forward five years, you'll be expected to have the capability to add or delete network nodes and users at a moment's notice.

Figure 4. Communications questions about business technology and management

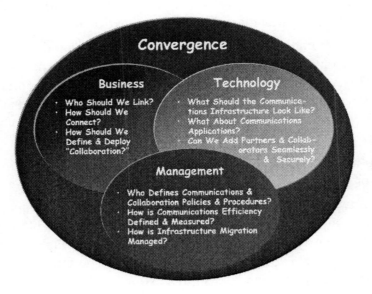

The whole concept of "communications support" is already obsolete, since communications is not a supporting player but an integral part of every aspect of your business process.

What About Transactions?

There's a pretty good chance that your applications portfolio is a hodge-podge of applications developed over the past 30 years or so that require some form of life support to exist. A depressing exercise would be to develop a pie chart that segments what you spend on applications life support versus strategic positioning through new application deployment. If I were running a business today, I'd want to see the trends here. One of the approaches to Y2K was to make old systems compliant. While that may have made sense, before taking that step I would have insisted on seeing the life support data. I would not have spent money on an application that was already costing me profit.

If you're in a large company, you've probably got some mainframe, client/server and Internet/Web applications (that are driving your e-business strategy) – and herein lies the problem: for the past 30 years we've been defining applications around silos and fiefdoms. In fairness, we developed applications around tasks we needed to complete. Initially these tasks were computational; over time, they became transactional, and now they're collaborative. Unfortunately, many of us are still just "computing." It's no longer about disembodied tasks, silos or fiefdoms. If you've got people that see the business technology world through adversarial glasses you need to take them out of the game.

The applications end-game consists of a set of inter-related, interoperable back-office, front-office, virtual-office, desktop, laptop and personal digital assistant (PDA and other thin client) applications that support collaborative business strategies. Put another way, continuous transactions that connect customers, suppliers, employees and partners are the lifeblood of your 21st century success.

A key application question should focus on the relationship between transactions and profit.

Do you know which ones yield the greatest profit? Do your applications facilitate the touching of employees, suppliers, customers and partners? How many applications do you have (that run on your desktops, laptops, PDAs and other access devices)? Do you have an applications portfolio management system that helps you locate and support your applications? How do you

Figure 5. Applications questions about business, technology and management

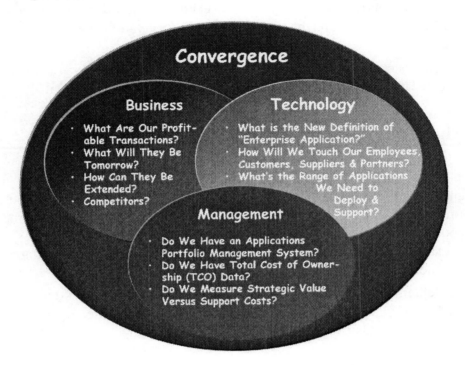

support all of these applications? Do you have in-house support – your own technology staff – or do you use a vendor to support your applications? If you outsource support, how well is the vendor performing?

Figure 5 lists some of the questions begging for immediate answers.

And Data?

Data is the lifeblood of your applications and your need to link employees, customers, suppliers and partners in a virtual world. We now think beyond database administration and about intelligent decision support, online analytical processing, data warehousing, data mining, business analytics, metadata and universal data access. Or at least we should. We should stop thinking about these technologies and tools as technologies and tools, but as integral parts of new business models like customization, personalization, up-selling and cross-selling.

The business questions about data include: can we cross-sell and up-sell? Can we connect everyone? Can we extend our business models through integrated data, content and knowledge bases? The technology questions address variation and integration, and the management questions address administration efficiency.

Figure 6 lists some of these questions, questions that should get you thinking about the inter-relationships among business, data and management.

Who Owns Security, Privacy And Trust?

Security – and its first cousin, privacy – are now household requirements. If you ignore them, you're toast. How did this happen so fast? Blame it on distributed computing – and the distributed steroid known as the Internet. As business models moved into cyberspace we found ourselves facing new threats. We're now surrounded by security and privacy technologies, officers, consultants and regulators.

The September 11, 2001 wake-up call also helped everyone focus on security. For lots of years, it was hard to get physical or digital security budgets approved. It's easy now.

Figure 6. Data questions about business, technology and management

Convergence

Business
- Can We Cross-Sell?
- Can We Up-Sell?
- Do Our Customer, Employee, Supplier & Partner Data Bases Integrate?
- Can We Extend Our Business Model?

Technology
- Do We Have Too Many DBMS Platforms?
- Do We Have Data Integration Capabilities?
- Do We Have Data Warehousing & Mining Expertise?

Management
- How Efficient is Our DBMS Administration?
- How Integrated is Sales, Marketing & Service?
- What Data Integration Opportunities are We Missing?

Trust is critical here. While many consumers have increased their online purchasing, there are still lots that have reservations about making serious purchases over the Web. Business professionals feel the same way about large digital business-to-business (B2B) transactions, and problems with spam and pornography continue to grow.

Denial of service attacks, viruses, sabotage and full-blown information warfare are all likely to increase as our dependency on digital transaction processing increases. Should I say this again? Yes: denial of service attacks, viruses, sabotage and full-blown information warfare are all likely to increase as our dependency on digital transaction processing increases. This is a perfect linear relationship.

Figure 7 lists some of the key questions.

Technology has to provide trust and protection in cost-effective ways, and all of the trust, protection and technology pieces have to work together in an environment that's procedural and disciplined. The key point? Trust and protection are business technology management goals, not technology goals.

Figure 7. Security and privacy questions about business, technology and management

How About Variation?

Variation in your environment – whether it appears in furniture, heating and air conditioning, your fleet or your technology infrastructure – is expensive. But the whole area of standards is fraught with emotion. Nearly everyone in your organization will have an opinion about what the company should do about operating systems, applications, hardware, software acquisition, services and even system development life cycles. Everyone. Even the people who have nothing to do with maintaining your computing and communications environment will have strong opinions about when everyone should move to the next version of Microsoft Office. In fact, discussions about standards often take on epic proportions with otherwise sane professionals threatening to fall on their swords if the organization doesn't move to the newest version of Windows (or Notes, or Exchange – or whatever).

What does management really want here? Most businesses don't associate standards-setting with business models, processes, profits or losses. Whether the environment has one, five or 20 word processing systems, variation is seldom associated with business performance since it's hard to link homogeneity with sales. But the fact remains that expenses are clearly related to sales, and standards are closely related to expenses. Herein lies the subtlety of standards and 21st century business technology convergence.

What else do you want? You want flexibility – and here lies the only sometimes-valid argument against standards. If your environment doesn't support the business computing or communications processes the business feels it needs to compete, there will be loud complaints. Business managers want to compute and communicate competitively. Standards are often perceived as obstacles, not enablers. But, almost always, nothing could be farther from the truth.

If we've learned anything over the past few decades, it's that standards are as much about organizational structures, processes and cultures as they are about technology. The ability to actually control computing and communications environments through thoughtful governance policies and procedures will determine how standardized organizations become. We've also learned that the more you succeed, the less you pay.

The other side of standards story is technology. Will the world migrate to Java applications or will extensible mark-up language (XML) obviate the need for common applications architectures? Will fast Ethernet grow dramatically? Will Bluetooth or other wireless standards like 802.11 dominate mobile computing? Oh yeah, this can get really annoying. If you find yourself at a meeting where debates break out about this stuff at this level of detail you need to either leave the meeting or immediately revise the agenda to kill discussions with no hope of ending. But this doesn't mean that the conversations about specific technology standards aren't important. They're enormously important. They determine how much business agility you have, how much business technology efficiency you enjoy, and how much you spend to keep the trains running on time.

But here's the key point: you're not in the technology standards setting business. Regardless of how brilliant your CIO or CTO is, he or she will have minimal impact on the ultimate direction the industry takes. (I apologize to the CIOs and CTOs out there, but it is what it is.)

The way to sidestep the stupid debates is to look to the major vendors and their commitments to standards. The vendors? The enemy? Yes. They – along with their consultant compatriots and the largest vertical industry players – control the direction and pace of standards. All you have to do is watch them, avoid early standards commitments, and then make them make good on their migration (to new standards) or integration (of older standards) promises. Making bets on the outcome of technology standards wars is dumb, especially when the costs of losing are huge and the probability of influencing the outcome is small.

Figure 8. Standards questions about business, technology and management

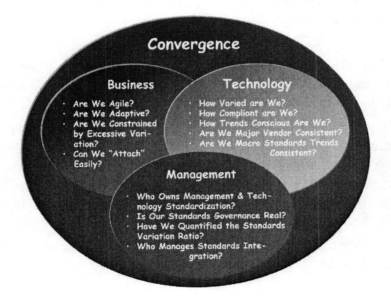

It's likely you've heard references to return on investment (ROI) and the total cost of ownership (TCO) every time the subject of standards comes up. Lest there be no misunderstanding here, there is no question that environments with less-rather-than-more variation will save money. Or put another way, you have some choices here. You can aspire to be sane or insane.

Figure 8 lists some questions that will help you implement a standards strategy.

So How Well Are You Organized?

Beyond the endless discussions about death-march CIOs who report to CFOs, and Cheshire CIOs who've landed seats at the big table courtesy of their CEO-reporting relationship, are huge issues around how to make business technology "work" in your company. Better yet, the objective is to have business and technology integrate in your company.

It's time to completely re-think all this.

For now at least, in a hierarchical organizational structure, your CIO and CTO should report to the CEO (later we'll talk about whether you still need a CIO or CTO). But this whole notion of technology people versus business people is obsolete. How can a technologist who's clueless about business be effective? How can a brilliant business strategist who's clueless about technology be effective? Give me a break: it doesn't get more basic that this. Prior to the start of the 2nd digital revolution it didn't matter all that much if there was no cross-fertilization, but now it determines market share.

Figure 9 lists some of the key organization questions.

The questions focus on the role that business technology plays in the company as well as the tools necessary to manage business technology assets. As always, incentives should play a pivotal role in business technology optimization.

We all know that it's naive to believe that behavior will change by redrawing organizational boundaries or by codifying new responsibilities. It always has – and always will be – about people. I don't care how many balanced scorecards you have, how many gurus you have under contract, or how large your market is, without the right people you won't succeed. They need the right skills and incentives, among other attributes.

In order to make 21st century business technology convergence work, several things about your people must be true:

Figure 9. Organization questions about business, technology and management

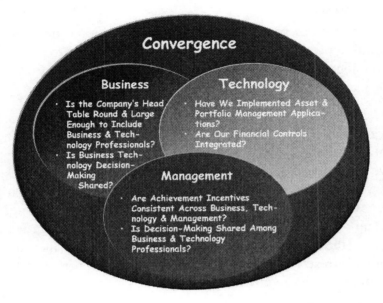

- Business skillsets must be re-examined. Skillsets that assumed discrete (versus continuous) transactions, relatively slow business cycles, limited supply chains, tightly managed partners, and mass marketing (versus mass customization), among other 20th century business models and processes, must be complemented by those who think about continuous, customized transactions in ever-collapsing supply chains and collaborative partnerships.
- Technology skillsets must also be re-examined. Skillsets that supported mainframe-based applications, data center operations and extensive software development are less valuable today – and will certainly be less so in the future – than architecture design, application integration, distributed applications (so-called network centric applications), project management and program management skillsets.
- Incentives must be re-examined. We must revisit the reward structure to make certain that the skills, talent and activities that mean the most to the company are generously rewarded, while those of less importance are rewarded accordingly. It's essential that the "right" message be sent here.

Employees must believe that there's a clear vision for the business technology relationship and they'll be rewarded for their commitment to this relationship.

- A new breed of business technology professional must be fielded, including professionals with an understanding of broad and specific technology trends, collaborative business trends, and how to convert the intersection into profitable business technology models.

This is very tough conversation. We'll really get into it in Chapter VII, but suffice it say here that there's enormous leverage in your courage to make objective decisions about people. Clubby, good ol' boy cultures, especially given the velocity of 21st century business, will become major drags on business effectiveness. Some industries will have to make fundamental changes in how they staff up – and out. Draw a straight line between profitable growth in your company and the individuals directly responsible for it, and then draw dotted lines to all of the rest. The solid, bold lines identify the keepers. The dotted lines identify some decisions you need to make.

Take a look at Figure 10. Many of the serious people questions are there.

Figure 10. People questions about business, technology and management

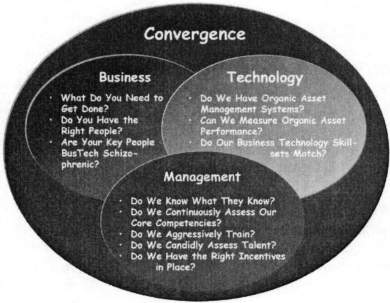

Who Owns The Future In Your Company?

Who's in charge of tracking business technology trends in your company? Lots of places have in-house gurus, but very few have created formal positions to track the major business technology trends that can impact their companies. I must confess that I've always found this amazing given the pace of business technology change. Maybe it's time for all of us to rethink our business technology watch strategies.

So how do you identify the business models and computing and communications technologies most likely to keep your company growing and profitable? The explosion in technology has changed the way you buy and apply technology and has forever changed expectations about how technology can and should influence your connectivity to customers, suppliers and employees.

What you need is a technology investment agenda that helps you identify the business models and processes and enabling technologies in which you should invest more and those that get little or none of your financial attention.

The agenda ultimately must be practical. While blue sky projects can be lots of fun (especially for those who conduct them), management must find the

Figure 11. How to think about business technology trends

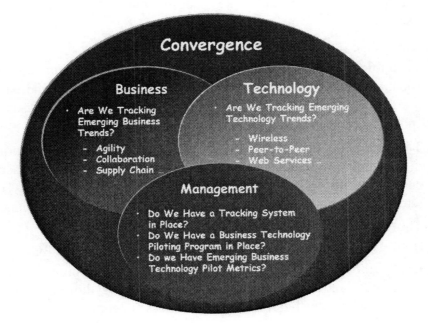

business models and technologies likely to yield the most growth and profitability, not the coolest write-up in a trade publication. But this can be tough especially when there's so much technology to track – and relatively little certainty about what your business models and processes will look like in two or three years.

We need to rethink strategic planning, scenario development and other activities that can reduce uncertainty about future business technology success. Figure 11 introduces some of the key questions about trends spotting.

Early Takeaways

This chapter has only scratched the surface of how we need to think about business technology convergence and what you need to know and do to kill the competition in the 21st century. I'm focusing here on business technology convergence because of the leverage one can gain through convergence and because of the money we can so easily waste if the wrong decisions are made.

This book is about the relationship among business, technology and management. Ideally, after reading it you'll **think** a little differently, **know** some things you didn't know before, and be able to **do** some new things.

Here's a summary of some of the key ideas from the above and following conversations:

- Technology decisions cannot be made in a business or management vacuum. All technology decisions touch internal (employees) and external (customers, suppliers, partners) players, and all technology investments should be driven by holistic strategies.
- There are management processes that can make business technology investments more cost-effective, processes such as **reasonable** business case development, assessments of total cost of ownership (TCO) and return-on-investment, performance metrics management and due diligence.
- The relationship among business, technology and management is inseparable – no matter how hard we try to treat them independently.
- Emerging business models will require collaboration, supply chain integration, agility and continuous transaction processing; the distinctions we now draw between business and e-business are gone.

- There's a range of **mainstream** technologies necessary to define and support successful existing and future business models, such as communications, applications, data and security technologies distributed across the enterprise.
- There's a range of **emerging** technologies, like wireless, Web services, natural language understanding and automation (including the "semantic" Web) likely to impact business the most.
- The range of organizational and people (political) strategies necessary to make all of the pieces work together including strategies like decentralization, standards-setting, project management and e-learning will require you to make some tough decisions about some good ol' boys.

Convergence Excellence

The greatest challenge – of course and always – is changing the way we think about the business technology relationship, and then dealing with the implications of really new ideas. Chances are you probably don't think about business and technology holistically, that your CIO still at least partially reports to the CFO (or not to the CEO) and that questions about technology tend to be expense-related, not strategic. There's lots to do but this time we cannot rely on the normal delays that have explained the evolution of business technology over the last 30 years: within three to five years all of this stuff will work together, your business model will have morphed horizontally and vertically and your key people will either be integrating seamlessly in your organization or doing it somewhere else. Don't be lulled by everyone's over-reaction to the bursting of the dot.com bubble. What's happening here is fundamental. Just when everyone thinks that business technology is business-as-usual, mega changes are occurring in what's possible, and in how business and technology will converge.

The key is to recognize the crossroads we're at now and begin to take steps to think about which new directions to take. Start with a business technology management health check framed by the questions posed in the Figures 2 through 11. The answers to these questions might help you better understand where you are now and where you should be going. We'll dig much deeper in the following conversations.

Does This Make Any Sense?

So what do you think?

The CEO …

"So basically you think we've screwed up the relationship between business and technology, and now we have to rethink it – again … haven't we been here before?"

The CFO …

"Sounds like 're-thinking' will cost money … will the outcome of all this require me to write yet another technology check? I remember everyone talking about e-business the same way … why is 'convergence' any different? I must say though that I like that question about 'profitable transactions' … do we know exactly where we make and lose money and what buttons we could push to make more?"

The CIO …

"Yeah, and what about all the good stuff we do day in and day out? Does anyone get any credit for this? The lines of business are clueless about what we do all day and have never given us clear guidance about what to buy – and not buy … all they want is everything to work all of the time and cost less each year … right …"

The CEO …

"OK, you deserve some credit … feel better? But I have to tell you that there's still a price on your head … there are a lot of people in the trenches that want you gone … I get calls every day about systems that crash … not to mention what your team costs all of us …"

The CTO …

"There's no justice … you want five 9s for no pain … our infrastructures and architectures are obsolete because you bastards won't spend any money on the basics …"

The Chief Operating Officer ...

"What the hell is an 'architecture'?" "What are 'five 9s'?"

The CSO ...

"All of this makes sense, I guess, but we still have holes in our infrastructure and environment ... we can get hacked anytime ... we need to spend a lot more money here ... I've been telling you this for years ..."

The CMO ...

"Is there a reason why I should be here? I don't hear anything really new ... you've been talking about all this for at least a couple of decades and nothing really changes ... if there's no new message here or something I can really spin, then I have some other things to do ..."

The General Counsel ...

"Me too ..."

The CEO ...

"Everyone, just sit down ..."

The Facilitator ...

*"Great group we have here ... pretty typical stuff ... but you've all missed the point. Here's what we need to do ... first, forget about Y2K, the dot.com bubble and the great deals you got on unproven technology and service models ... second, strip away the hype and remember that vendors and consultants are not our friends, at least not yet ... next, rethink your business models and the best that technology can offer, especially as it involves **collaboration and integration** ... I have to show you a slide ... look at this ... this is the essence of how you need to think – and what you need to do ... collaboration is about business and integration is about technology ... all of your business technology decisions should pass through the collaboration/integration filter ... **every God damned one of them** ...*

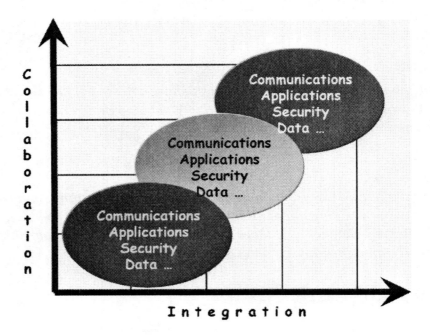

Stay out of the red zone, minimize your time in the yellow zone and work the green zone to death ... this is where you want to live ... it's that simple ... if you have business models that are not already collaborative or not heading fast in that direction, kill them, and if you have technology that doesn't work together today and isn't likely to tomorrow, kill that too ... okay?"

The CFO ...

"I like this picture ... finally, a hammer ..."

The CEO ...

"I like it too, but it's a hell of a lot more than a hammer ... it sounds like a way to finally get all this stuff to work together ... good timing, by the way, because I'm tired of dealing with the 'relationship' the same old way ... and the board of directors is actually starting to understand this stuff ..."

The CLO ...

"Did someone give them 'The Idiot's Guide to Technology'?"

The Facilitator ...

"Cute ... let's continue with a conversation about what collaboration really means ... one step at a time – but I guarantee all of you that this will work – at least a whole lot better than things work now ... stay with me ..."

The COO ...

"Not so fast ... summarize all this in English ..."

The Facilitator ...

"Fine ... for several decades the relationship between business and technology evolved as silos ... there were people responsible for business, technology and the management of all this stuff ... look at the next picture ...

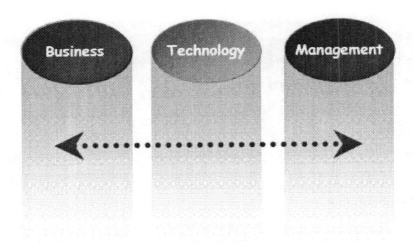

... the arrow suggests that we tried to 'align' these activities but by and large we failed ... why do I say that? Look at the next slide (I know you, COO, you love Powerpoint slides) ... the silos were disconnected (as silos

often are), we didn't really leverage technology: we developed business strategies and then turned to technology to make it all work, under-emphasizing the role of technology – just as, by the way, business planners under-appreciated what technology could really do for business ... our management consisted of a set of 'worst practices' that we repeated for decades, practices that were exploited by vendors and consultants ... and to make matters worse, we confused ourselves about how all this should be organized ... the net effect is probably the worst track record – 75% probability of failure – in the history of business ..."

20th Century Alignment

- Disconnected Silos
- Sequential Business → Technology Acquisition & Deployment Processes
- Business Strategic Planning Under-emphasized the Role of Technology
- Technology Under-Appreciated Its Role as a Business Enabler
- Poor Management Practices Exploited by Vendors & Consultants
- Confused Organizational Roles & Inadequate Governance
- Horrible Success Record ...

The COO ...

"Maybe I didn't really want it summarized ..."

The Facilitator ...

"It gets better ... all of this was exacerbated by the perfect storm ... our temporary insanity over Y2K, e-business and how cheap the VCs were selling killer apps ..."

The COO ...

"Give it to me straight ... the whole truth and nothing but the truth ..."

The CEO ...
*"The **truth**? You can't handle the truth ..."*

The CIO ...
"Let's get to the good stuff ... so what else are you telling us?"

The Facilitator ...
"The future is about business technology convergence ... the silos have to come down ... we have to recognize that business, technology and management are inseparable and that this inseparability has huge implications for how we develop business technology strategies, how we buy stuff, how we make it work together and – essentially – how we kick the competition's ass ... look at the next slide ...

21st Century Convergence

- **Inseparability of Business, Technology & Management**
- **Implications of This Inseparability**
    - Business Trends + Technology Trends + Management Best Practices = Convergence
    - Changing Organizational Roles & Responsibilities
    - All Guided by the "Right" Questions About Business Technology Convergence ...

The CEO ...
"This is starting to make some sense ... tell me more about these 'implications' ...

The Facilitator ...

"Sure ... business is changing ... all this 'collaboration' talk is real ...here's what the new business priorities and capabilities look like ...

... and here's what the new technology priorities look like...

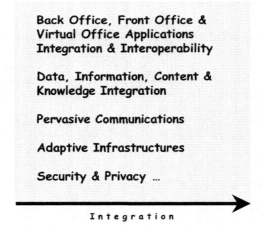

... and here are the management priorities"

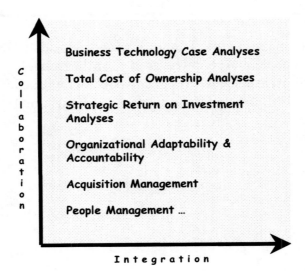

The CEO ...
"So we have some work to do ... and all those questions you just asked need to be answered before going forward – before spending another hundred million dollars ..."

The CFO ...
"Halleluiah! I've died and gone to heaven ..."

The Facilitator ...
"Not quite yet ... now we need to drill down on all these ideas ..."

The CMO ...
"I'll be back ... I just need to make a quick call ..."

Endnotes

[1] Paul Strassmann's work is impressive and persistent. He's published a number of books and countless articles on the technology/productivity relationship. Go to www.strassmann.com for more details, access to his articles and information about his books and consulting services.

Chapter III

The Business Conversation - Where We're Going

So what do we have here? This conversation will focus on:

- **Collaboration** and all of its flavors, including supply chain management, personalization, customization, optimization, automation and trust – and how if you're not thinking about collaborative business models you're toast.
- **Continuous versus discrete transactions** – the big change from the 20th century which, among other things, has obliterated the distinction between business and e-business and forced us to think about "whole customer/supplier/employee/partner management."
- **Real-time analytics** – your next best friend.
- The collaboration and integration investment filters as **uber-filters**: if business technology investments don't pass, you don't do them.
- **Business scenarios** and how to build them, because you have to model your collaborative future before you can make the right business technology investments.

So What Do You Think?

Enough of the preliminary stuff – let's get started.

You make things, you sell things, you service things. Ten to 15 years ago the processes that optimized production, sales and service were relatively well-bounded, with clear beginnings, middles and ends. Today those processes should be extended and continuous. This is the essence of what's happened to business over the past 30 years and what technology's made possible. This is the change that's so essential to understand, this is what technology enables. One cannot exist without the other. This conversation will focus on emerging business models. Chapters IV, V, VI and VII will identify the technologies and management best practices that should converge with the new business models to complete, to win. You see, this is at the core of the whole business technology convergence message: holistic thinking about business and technology wrapped in solid management practices. This is the new core competency. The anti-competency is linear, sequential thinking about business/technology alignment.

Sounds great, but what the hell does it mean? One way to think about the change is to contrast discrete transactions with continuous ones. Discrete transactions – like selling insurance policies, buying disk drives to be included in PC manufacturing, or buying or selling stock – used to be discrete transactions that one could begin in the morning and compete by afternoon – or the next morning after a good night's sleep. Even today these transactions are continuous. One insurance policy is blended with another. Cross- and up-selling are continuous goals. Disk drive acquisition is integrated into PC manufacturing and buying and selling stock is simultaneously linked to tax calculators and estate planners. Tomorrow all of these transactions will be extended, continuous **and** automated.

Another way to think about all this is to imagine what would be possible if your company had immediate and continuous access to its employees, customers, suppliers and partners. What if you could communicate with all of these constituencies whenever you wanted? What if you could tell them about new deals, discounts and opportunities anytime at all?

What if you could connect all of your suppliers with your manufacturers, marketers, salespersons and customers? In spite of what supply chain vendors tell us today about how integrated supply chains really are, they're only partially connected. But within a few years they'll collapse into webs of efficiency that will blow away anything we've imagined to date. Technology investments today must anticipate the connected tomorrow.

Five important questions:

- Do you think about business this way, as a continuous process of action and reaction that will eventually become quasi-automated?
- Do you think about "whole customer management," where customers have life cycles that can be monetized?
- Do you think about supply chains that connect manufacturing with inventory control, distribution and service?
- Do you think about how transactions can be automated?
- Do you filter your business and technology decisions with a collaborative mindset?

What do you think? Can we get there from here? One thing's for sure. If we continue to think about business processes as discrete we'll miss profitable opportunities for continuous transaction processing. Worse, if you keep making investments in plant, equipment, communications, data bases and applications that are inconsistent with the collaborative, continuous world, you'll waste a ton of money positioning yourself to succeed in the past.

We're at a very different place now. Do you see it? Connectivity among employees, suppliers, customers and partners – though far from complete – is enabling interactive customer relationships, integrated supply chains and the business analytics that permit real-time tinkering with inventory, distribution and pricing. The strategies and business models were always there. For years we've imagined seamless connectivity and ubiquitous business. But the convergence of business and technology has made it more than just possible.

The net effect is that time and distance have been compressed, and speed and agility have been accelerated. Some strategists get this and some don't, just as some warriors understood the impact of the longbow and some didn't. These are not debatable issues. If you have people in your organization that still challenge the importance of connectivity, collaboration and continuity then it's time to re-staff. (I remember not too many years ago when a senior guy in a company told me that the Internet was a "fad," and that it would disappear in a couple of years. The same company believed that employees should not have access to the Internet and that e-mail undermined productivity. If you find yourself surrounded by this kind of thinking – and you're in charge – then you need to whack the Nay Sayers and immediately get some objective counsel.)

The collapse of the dot.coms got way too many of us to believe that the new digital economy died before its time. The irony is that the hype clouded the preview of a classic movie set to be released in just a few short years. But the plot remains what it always was. Here's what you should assume:

- **The distinction between business and e-business is gone.** There is just **business** now with one additional channel, the World Wide Web. The total integration of off-line and online transactions is inevitable – and well on its way. **If you have "e-business" or "e-commerce" divisions or practices, fold them into existing, evolving initiatives.**
- The distinctions among business-to-business (B2B), business-to-consumer (B2C), business-to-employees (B2E) and business-to-government (B2G) are evaporating. **Well-bounded business models and processes are morphing into blended, continuous transactions.**
- **Collaboration** is the watchword for the 21st century, the umbrella under which supply chain planning and management, mass customization and personalization, dynamic pricing, and a whole host of new models and processes live. **Talk about it, invest to achieve it, and train around it.**
- **Eventually many old and new processes will be automated.** It makes no sense for a procurement officer to manually repeat purchases that occur month after month. Such management should be by exception only; once automation occurs business becomes continuous.
- **Every business and technology decision you make should be made through the collaborative lens that also sights integration.** Every glance backwards or sideways from this lens hampers your ability to compete.

The last bullet is at the core of business technology convergence. It's also a test. You can take it yourself and give it to your trusted advisors. You can grade it and weep. Why? Because in spite of how well your people have mastered the buzzwords, they probably have no idea what we're talking about here. The change in business is so profound that it's hard for people to absorb. The idea, for example, that one's manufacturing supply chain could be fed by real-time component quality data or that customer requirements could be personalized during assembly is still vague to many strategists – and lots of managers. The problem is that we're conditioned to think that change occurs gradually (especially after dispelling the myth of the Internet "killer app") and we discount the importance of the impact of multiple events, opportunities and technologies. We do this because it's so hard to conceptualize things holistically and infer the impact on your business and your competitors' business. How candid should we be here? Executives and managers that grew up with discrete transactions and back room "support" technology have a very difficult time understanding all this. Yes, they get some of the lexicon but they cannot easily comprehend its significance. (By the way, where do you get your "radical" ideas? Who thinks outside-the-box in your company? Are they rewarded – by you – or are they forever branded as off-the-wall? Who's your collaboration guru?)

With the gloves completely off, I'd argue that half your team is pretty clueless about the new business models and how they converge with technology. They're also ill-equipped to implement evidence- (versus bias-) based management practices. The other half are only slightly aware of what's happening. There might be a few that fully understand how fundamentally things are changing, but most of them are struggling to comprehend how collaborative trends intersect with technology. Why is this so difficult? The answer is simple. It's because in order to fully appreciate what's happening, you have to know about vertical business trends, trends in computing and communications technology **and** creative management techniques. The leverage lies in convergence, but in order to find it you have to understand all sides of the equation. Very few people (and that includes consultants) really understand them all.

Figure 12. Collaboration drivers of business technology convergence

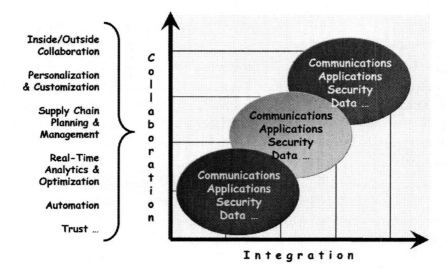

There are a handful that have grown up in technology that have morphed into business strategists and vice versa, but by and large, people that understand the convergence of business and technology are few and far between (if you have any, make sure to keep them happy).

Let's dig a little deeper into what's happening. Remember the picture in Figure 12? We're now talking about the left side and what it means to identify collaboration as the business trend you should understand.

Inside/Outside Collaboration

The idea here is simple: strategically connect employees, customers, suppliers and partners for competitive advantage. Two things have to be true to make collaboration work. First, collaborative processes have to be defined. For example, Dell assembles computers by exploiting an integrated supply chain that consists of suppliers, employees, customers and partners. All players have a role to play. Suppliers provide stuff that gets assembled while employees monitor the quality of the stuff and the assembly process. If, for example, a batch of disk drives under-performs then a request for supplies is broadcast to alternative suppliers. Customers can customize their computers online where they can also buy software and printers (and other devices) from partners. The process defines collaboration – technology enables it. While it's all about collaboration, the process has to be defined, efficient and profitable. In other words, it's possible to define inefficient and unprofitable collaborative processes, just as it's possible to deploy the wrong technology (to enable the wrong collaborative model).

The trick is to define collaborative processes that improve efficiency, reduce cost, improve profitability, increase market share, extend the staying power of Viagra and bring peace to the Middle East. Let's keep it simple: you need to **define** collaborative business models and then **deploy** technology that enables profitable growth (but as we'll discuss in Chapter IV, that technology has to integrate).

Our first steps toward employer/employee collaboration were enabled by e-mail and other workflow applications, but the story got interesting when we launched Intranets which were early internal portals to applications, data, news, benefits, operational processes and communication. Many companies built Intranets to solve lots of internal collaboration problems (and software vendors subsequently exploited the trend with knowledge management and off-the-shelf enterprise portal applications). But what's the point? Why should anyone update a benefits package through a mass mailing? Why would we ever announce personnel policy changes with paper memoranda? Instant updates, immediate communication and continuous documentation are all enabled with simple Intranet technology. It's fast and cheap.

Take the idea and extend it to your customers. Touch them digitally. Obviously this does not mean that you will **only** touch them digitally, but that digital contact represents another way to stimulate and serve them. Convenience – from your customers' perspective – is the reward here. From your perspective, it's cost-effectiveness and a way to extend your communication with your customers, communication you can use to please them, up-sell them and cross-sell them. (But be a little careful here. Lots of technologists can promise lower customer service costs by off-loading service to the Web, where a 24/7 Web site can answer lots of "frequently asked questions." But lots of your customers hate browsing through layers of content to find what they need. Lots of people still want to talk to another human being. It's important to balance organic and digital contact, and not "punish" your customers, employees, suppliers or partners with digital-only access to important information.)

Now take the collaboration idea and extend it to your suppliers. While we'll talk more about supply chain planning and management in a moment, supplier collaboration can take various forms. For example, do you plan with your suppliers? Do your suppliers have a window into your production processes – if you co-produce – or into your demand for their products if you're a wholesaler or retailer? Do you have multiple suppliers and do they all collaborate?

Last but not least, extend the idea to your partners. Do you sell your own stuff as well as stuff produced by others? Do you sub-contract products or services? Imagine a capability that would enable you to plan with your suppliers and partners to facilitate collaborative forecasting and planning. Imagine a capability that would reduce transaction friction, excess inventories and pricing instability?

Collaboration tightens the value chain of your business. Thinking this way is a derivative core competency. If your executives and managers aren't thinking about the costs and benefits of collaboration engineering, then you have a problem.

All of this connectivity goes under lots of names these days. I've selected collaboration as the general term that conveys inter-connectivity among employees, customers, suppliers and partners. Jim Champy calls it X-engineering – another good term because it implies that business processes need to be extended in multiple directions simultaneously.

We'll look at a number of flavors of collaboration you need to try. As always, remember the decision filters: if applications, infrastructure, communications and security investments aren't consistent with collaboration (and integration) then you need to ask some tough questions – and then shoot the anti-collaboration/integration planners.

Supply Chain Planning and Management

This is the classic collaboration dream. It's also one that's ripe with clichés. The fact is that when you say "collaboration" many people say "supply chain." Is this a good thing or a bad thing? Well, first it means that supply chain planning and management are on the top of everyone's list of important things to, or at least think about, even if many of the people thinking about it really don't understand it.

Supply chain planning, management and optimization is a subset of collaboration and of course enables lots of things including customization, personalization, dynamic pricing and automated transaction processing. Everyone has supply chains. Even companies whose principal products are ideas. Understanding where the leverage is and how it can be optimized in your supply chain is essential to successful collaboration. Let's begin with a look at the production of ideas, since that's the counter-intuitive understanding of supply chain planning, management and optimization.

Good ideas are produced by a few smart people. The raw ideas get packaged by any number of clever people (yes, it's easier to find clever people than smart people). Anal people then price the idea (no comment on the value or availability of anal people). So how do you plan, manage and optimize this process? Let's assume that you're in the research and development business. You have some solid producers and some mediocre ones. You have defined processes for packaging ideas, processes for assessing ideas, and even processes for sharing ideas with partners and prospective distributors. If you plan properly, your supply of good ideas will remain constant, your packagers efficient and your pricers anal. Optimization happens when you increase the number of monetizable ideas. You do this by increasing the number of smart people and working the clever and anal ones harder. What? Think about it: optimization is primarily about the number and quality of ideas and the capacity to handle them. If the number of ideas rises too fast then you'll have to increase the number of clever and anal people, but they're easier to find and keep happy than really smart people.

Let's switch to manufacturing. As we all know, complex production processes require enormous amounts of coordination. Digital technology can help coordinate complex, distributed manufacturing processes. **Complex** and **distributed** are the key words here. As we outsource more and more component manufacturing, the need for digital glue is rising dramatically.

But the supply chain mantra is much broader, and while we're a few years away from completely integrated supply chains it's time to start thinking holistically – **now**. The leverage here is huge. Think about it as a kind of controlled, vested-interest-based partnership among all of the stakeholders in the value chain or, put more crudely, all of the people with which you deal who want to make lots of money. Partnership here is based on a little skepticism and lots and lots of shared communication.

The potential efficiencies and cost-savings are enormous – **if** value and supply chains are mapped accurately and integrated effectively. The mapping is critical. Have you modeled your value chain? Do you know the ingredients to your profit pie (chart)? One of the earliest questions raised here is about profitable transactions. Can you identify them quickly? And where do you lose money? Value chains inform supply chains which can be extended as far and wide as you wish, so long as your leverage remains intact. In other words, while it's possible to map your company's entire value and supply chains – of course in an exquisite Powerpoint presentation – it's impossible to exert influence on your entire extended supply chain. Influence will range from total control to "won't you please consider joining our consortium?"

The pieces include supply chain planning, supplier management, demand forecasting and the analytics to make it all work. There are process and technology standards to watch here as well. Your supply chain is affected by these standards and you therefore cannot ignore them. An important point: connections occur perceptually, procedurally and technically. All three must exist for supply chain planning and management to actually work. Perceptually, the members of supply chain need to want to collaborate or, put a little differently, there needs to be clear vested interests in their participation. If they don't feel it makes top line/cost management/bottom line financial sense then they won't play – or at least they won't say they won't play, but their participation will be slow and cumbersome. Never underestimate the difficulty of selling "whole-is-greater-than-the-sum-of-the-parts" arguments. Lots of people believe that collaborative teams – supply chains – only work for a few of the "partners." "Competitive partnering" is sometimes the best way to understand how supply chains really work. In fact, there are those who believe that there are clear supply chain hierarchies where some members participate because they have no choice. Dell and HP, for example, had a good partnership going, where Dell resold HP printers, but after HP and Compaq merged, Dell announced that it would private label its own printers. HP quickly announced that it would no longer channel its printers through Dell. The fragile relationship between the two companies cratered when the vested interests changed.

Where are you on the supply chain continuum? Are you a leader or a follower? Depending on the answer, you'll have to plan accordingly.

Procedurally, supply chain planning and management requires a reciprocating process. Rules have to be established and followed. Technology vendors have developed software applications that manage the process according to sets of rules specified by the collaborators. The perfect supply chain of course is one where everyone agrees on the processes and everyone then follows the rules. Procedural mapping is critical to success and begins with a mapping of your current "manual" or automated supply chain. Here are some basic questions:

- How many suppliers do you have?
- How many buying processes do you support?
- What's your cost structure look like?
- Which suppliers cost the most, deal the most and argue the most?
- Do you know where demand comes from and how it changes over time?
- Can you predict demand?
- Is demand forecasting collaborative?
- Are new products collaboratively designed?
- Is order management and fulfillment integrated across your supply chain?
- Do you manage and optimize supply chain events (transactions)?

If you can answer these kinds of questions, you're in pretty good shape – especially if the answers are accessible through an interactive, decision-support system tied to your suppliers, partners and customers.

Collaborative planning, forecasting and replenishment (CPFR) alone can save tons of cash. CPFR is the essence of inventory management and leads to improvements in forecasting, sales and services – **when it's done right**.

Technically, there are lots of steps you can take – after you've adopted a supply chain mindset and **after** you've defined your supply chain planning and management processes and procedures. Everyone knows about the enterprise supply chain software vendors like i2 and Manugistics, but other enterprise resource planning (ERP) vendors also offer supply chain software – not to mention lots of smaller niche players. When we talk later about standards in Chapter IV, you'll see how decisions about business models and technology standards converge in the supply chain area, as they do in lots of other business technology areas. You've also heard lots about electronic data interchange (EDI) and other technologies like extensible mark-up language (XML) and business process interfaces (BPIs) that make the inter-connections possible. In addition to access, collaborative partners need connections to ERP data, procurement and billing applications, "storefronts" and full-blown exchanges, and sometimes custom connections with selected partners through virtual private networks (VPNs). Yeah, it's complicated (but awesome, when it all works).

What's The Reality Here? If This Is A 10-Year Process, What Year Is It?

We're in year six. Some vertical industries are in year three and some in seven. It nets out at about a six, which means that significant progress has been made but lots of work has yet to get done. CPFR is, for example, is a process standard that companies are embracing at a deliberate pace. Towers of Babel are slow to crumble – or integrate, and larger capital market trends will likely predict supply chain planning and management adoption rates. As supply chains get defined, the vested interests of the collaborators will get quantified, which in turn will permit companies to determine what supply chain role they want to play. Companies also need to locate themselves in the supply chain options space. There are relatively simple buyer/supplier relationships, private supply chain exchanges, full-blown vertical markets and even global supply networks. Will market-makers survive? In the late 1990s business-to-business (B2B) exchanges were all over the place. Companies could buy and sell chemicals, office supplies, auto parts and all sorts of things from private B2B exchanges – market-makers – that hosted buying and selling for a small transaction fee. But they by and large failed to sustain themselves for a variety of conceptual, procedural and technical reasons.[1]

Here's the deal: to make any form of collaboration work you have to synchronize all three parts. Obviously you should start with the concepts to see if your collaborators get the mindset. If the light's green you can define collaborative processes and procedures and test your supply chain to see if it has the right mindset (after you've defined what it means for the participating companies and what it does to their vested interests). Then **and only then** should you think about technology.

Customization and Personalization

Every so often I get an e-mail (or snail mail) from a company that's profiled me. It's analyzed data about where I live, what I earn and what I buy to determine what I like and what I'd pay for these products and services. Whenever I watch television I get bombarded by mass advertising that's usually lost on me. I can never remember the product or service, especially when the quality of the form and style of the message is good. That's one of the problems with mass marketing and advertising. The content of the message often gets lost in too clever forms designed to appeal to as many people as possible. As they say, half of all advertising is wasted, but no one ever knows which half. Ah, but that's the strength of mass customization through personalization: everything's pitched at me, a unique consumer.

Mass customization is cost-effective. It's built on much the same data that mass marketing assumes but the idea is to infer beyond the simpler correlations – like age, wealth, time of year – to specific ideas about what you and me would really like to buy, based on inferences about us as part of a larger group **and** as individual consumers. If you could reduce your mass marketing budgets and increase your sales more than the added cost of personalization you win the contact game. Sales rise because personalized contact is more targeted than mass marketing contact and therefore most of the communications are at least relevant.

You can "personalize" your contact with customers, supplier, partners and employees, and you can personalize all varieties of messages including sales, marketing, service and distribution. You can also personalize with paper, telemarketing, advertising and e-mail. Over time, given how low digital transaction costs are compared to other ways we touch the members of our value and supply chains, and how ubiquitous digital communication is becoming, it makes sense for you to reassess your budget allocation. You're paying for lots of channels to these people. Which ones pay the best?

There's a great scene in the Tom Cruise/Steven Spielberg film, "Minority Report." Cruise is walking in a city in 2054 and as his eyes are scanned he's immediately pitched a whole slate of personalized products and services. Imagine waiting for a plane or train and receiving countless messages about stuff you could buy that you already like and use all of the time – but now's on sale 12 feet away?

What if you could approach all of your employees, customers, suppliers and partners in ways that matched their interests, values and personalities? Remember that personalization and customization extends beyond customers and includes all of the stakeholders in your value chain.

How do you get there from here? How much data do you have on all of these characters? Forget customer focus groups: get to the empirical data that permits profiling. Invest in behavioral models that explain why your existing customers do what they do and what your future customers will value. Is privacy an issue? You bet it is. We're on a collision course here. Access to the kind of data we need to personalize and customize products and services is arguably very private data that's owned exclusively by individuals who should approve or reject requests for access to their empirical history. But even as privacy regulations and laws evolve in favor of consumers, there are still lots of clues out there that will fall beyond the scope of most privacy restrictions. Regardless of how this all plays out, it's essential that you develop the means to personalize and customize your products and services in ways that permit you to adapt to changes in privacy laws, your customers' behavior and distribution.

How do you do this? It's all about the depth, location and quality of your customer data. Does it exist? Who owns it? How good is it? How does it get better? It's also about the analyses you perform on this data. Some companies have excellent behavioral scientists who run the data every which way in search of correlations that explain what customers value and why they buy what they do. Years ago in graduate school I wondered why I had to learn all about multiple regression equations. Now I understand how sales and marketing professionals could not possibly survive without solving all sorts of equations designed to explain which factors (independent variables) explain which outcomes (dependent variables). The power here is amazing. It's possible, for example, to determine the following about individual customers:

- When to digitally (via browsers, cell phones, wireless PDAs, pagers) interrupt customers with a deal – and when not to interrupt them.
- What size discounts need to be, by person, by season of the year, by customer location and time of day.
- What combinations of products can be sold and what products don't mix with others.
- What short-term and longer-term life events influence which purchases.
- What forms and content of customer service each customer prefers.

This is just a sampling of the kind of personalized and customized inferences that can be made that will increase your sales and reduce your sales and marketing expenses.

There are some weird things that will happen along the way to successful digital personalization and customization. One of them is the annoyance factor. It's important that we validate our inferences before launching digital attacks on our customers. While I might like being offered a 33% discount on pizza if I stop in 30 seconds at one of my favorite restaurants, I might also find it annoying that Dominic knows where I am (through access to global positioning data), knows that it's lunch time and knows that I think his pizza is overpriced. My annoyance might actually evolve to anger as I piece together just how manipulated I feel.

Personalization and customization are serious sales and marketing weapons, but – like many powerful weapons – can be misused in the wrong hands.

Personalization and customization should not stop with your customers. The same analytical approaches we take to profiling customers can be used to profile employees, suppliers and partners. In fact, the personalization and customization of employees, suppliers, partners and customers will be part of the collaboration process. Plan for it.

Like all of the collaboration-driven business models, however, personalization and customization is first and foremost a process long before it's a technology. One of the reasons why companies continue to behave like crash dummies is that they continue to believe that technology is a remedy for bad business. Or, perhaps worse, they think technology will help a good business model that actually sucks. Personalization and customization is a state-of-mind about how to touch people and organizations. A good software package that discovers key inferences or facilitates the right communication at the right time is, of course, necessary to successful customization and personalization but it's not sufficient for successful contact. Your company's strategy and culture need to align with your personalization and customization objectives. Senior management must stay the course it sets here or all investments will be lost. This stuff only works when it's well conceived and well executed.

Let's dive a little deeper. Customers, suppliers, employees and partners all have "life cycles," which define how you monetize the relationships you maintain with them. The sales and marketing expression of "cradle-to-grave" pretty much summarizes the approach. The objective is to become a full-service partner with this diverse group for as long as possible.

In order to do this you need to collect as much data about each group as cost-effectively as possible. Unfortunately, and as we'll discuss later, you're already collecting lots of data: you just can't always get to it. The reason why there's a multi-billion dollar data integration cottage industry out there is because everyone wants to analyze data stored in different places, data that won't communicate unless lots of people spend lots of time and effort "extracting," "translating" and "loading" data into "data warehouses" designed as clearing houses for otherwise unfriendly data. We'll talk in the next chapter about how to avoid getting in data trouble but for now keep in mind that the objective is to get clean data in a place that permits you to ask all kinds of questions about your customers, employees, partners and suppliers.

If you get this data into shape you can analyze the hell out of it. The results will permit you to determine where expenses and profits are, how to manage marketing campaigns, and to conduct loyalty analyses among all sorts of other analytical insights and plans that translate to interaction with your constituents. This interaction can take the form of personalized marketing, sales and service through all sorts of contact media, such as direct (paper) mail, call centers, e-mail and advertising, which can occur in any number of physical and digital ways (stores, cell phones, PDAs, laptop computers, newspapers, etc.).

The key question is: what are you doing to enable "whole customer management"? What data are you collecting and analyzing to widen, deepen, manage and monetize contact with your value chain? If you're not sure, make a note to call someone about this.

Real-Time Analytics and Optimization

Think about real-time analytics – and the optimization of processes and outcomes that it enables – as your corporate dashboard. When you're driving your car there's a ton of active and passive data available to you from a glance or a command. Much of this data, like speed and revolutions per minute, is displayed continuously while other data's presented when something goes wrong. But what mechanical systems cannot easily do is conduct what-if analyses of seemingly disparate data to discover counter-intuitive correlations among various behaviors. You can't ask your car questions like: "If I drive you at 100 miles per hour for 11 hours on mountain roads while outside temperatures range from very cold to very hot, which systems are most likely to fail – and in what order?" But you can ask your customers – through the behavioral data you collect on them – what they like, when they like it, and what combinations of products and services make them happy. You can also ask technology questions. You can ask which applications – like your ERP, CRM and other enterprise applications – touch the most customers with the highest (or lowest) margin products or services. You can ask the systems that maintain your technology infrastructure about how well the infrastructure is performing, how much it costs to maintain the infrastructure and which technology within the infrastructure is about to break. Can you really do all this? Yes – if you invest in the right analytical applications and the right infrastructure technology.

Real-time insight into your business process and technology infrastructure, and the collaborative transactions they enable, will yield the flexibility you need to adapt to market opportunities and competitive pressures. Dashboard data's essential to maneuverability, because driving blind is dangerous. But think about this: we've been driving blind for decades. Twenty years ago we ran "batch" analyses on last quarter's sales and expenses. Ten years ago we had monthly results and turned the data over to sales and marketing for further analysis. Today we can change prices almost instantly but still can't "see" transactions, distribution or service. Nor can we immediately tweak our supply chains when some of the pieces need adjusting.

Analytics enable **business intelligence** (a term I'm sure you've heard over and over again). Like everything else, there's a process here. We'll talk more about data in the next chapter, but as we discussed during the personalization/customization conversation, the location and quality of data about your customers, employers, suppliers and partners determines just how far you can analyze business processes. If your data's all over the place, ugly, dirty and in any number of proprietary vendor silos, then you have a problem. Data warehousing vendors are of course happy to sell you software, gear and services to fix the problem. But it makes more sense to avoid the nasty integration process altogether.

Once the data's accessible, you can analyze it. You can query the data, develop reports, perform "online-analytical-processing"-based analyses, "mine" the data, visualize the data, export the analyses (to desktops, laptops and PDAs, for example) and ultimately use the data to make decisions – to rethink sales, to tweak the supply chain, to manage the back office (human resources, finance, accounting, manufacturing, etc.).

When we talk about "real-time" analytics we're talking about the ability to convert real-time analyses into immediate action. For example, let's say that the raincoats that you expected to sell were still sitting on the shelf. Your supply chain visibility permits you to see that the coats aren't moving. The same tools can help you predict what sales volume will look like over the next few weeks, given the drought conditions in most of the country. Given how unlikely it is that you'll sell the coats at the current price, you can roll out some price changes and then (almost) immediately calibrate sales impact. Predictive analyses can be enormously valuable to you as you make your way out of messes like this. And it can get pretty complicated involving hypothesis testing, pattern analyses, simulations and what-if sensitivity analyses, among other techniques. But keep in mind that predictive analyses are not just useful for averting disasters. They can be used to proactively optimize prices as well. For example, the real-time generation of demand curves can enable upward price adjustments for goods and services whose prices would otherwise remain static. Optimization software drives dynamic pricing, but lots of things have to be true for it to work as advertised. The supply chain mindset needs to be entrenched, data needs to be clean and accessible, and the technology to access and analyze the data must be reliable. Lots of vendors sell strategies and even more sell technologies embedded in large and small applications that slice and dice your (clean, available, organized) data in ways you can't even imagine.

More recently, analytics have become the anointed savior of the customer relationship management (CRM), enterprise resource planning (ERP) and e-business applications projects that went berserk. After many millions of dollars and tough questions about return-on-investment (ROI) – questions that were difficult if not impossible to answer – project champions, systems integrators and strategic consultants searched for ways to justify huge enterprise technology investments. Enter analytics, business intelligence, and optimization. The idea is simple enough. Integrate data from front-office applications – like sales force automation (SFA) and CRM applications – ebusiness applications, and back-office ERP and supply chain management (SCM) applications, and then pump it into a full-blown business intelligence environment that enables the analytics that trigger decision-making. Simple enough, huh? Well, the concept certainly is. But execution is a bitch.

So is it worth it? Absolutely. Especially if you're in the early stages of building a collaborative culture and you haven't yet invested tons of cash in a database/warehouse/mining/analytics infrastructure. But even if you've spent a ton on this stuff with anemic returns, there are ways to optimize the investments to achieve optimization through business intelligence. This is an area that demands investment, especially as the pace of change increases and we deploy more and more applications that must integrate.

Who owns analytics and optimization in your company?

Automation

Does anyone believe that customers, partners, employers or suppliers want to go to the Web to execute the same transactions day after day, week after week? It's goofy to think that a procurement officer wants to visit the same Web site every month to order the same number of 55 gallon drums of chlorine, or that undergraduate students want to search for Dave Matthews CDs week after week? What about your systems? Why can't their condition and effectiveness be automatically monitored – and fixed when a problem is detected? Your back-office systems can also automatically transact all sorts of business. All you need to do is engineer them to accept dynamic instructions. What's the difference between a "regular" computer program and one that's "smart?" The regular one computes and re-computers (over and over again) transactions it's been programmed to execute. There's not too much beyond these computations that it can do. But a "smart" application is capable of reacting to outside variables and then – through a pre-programmed set of rules – execute a different calculation each time the outside variables are different. Smart programs can monitor networks - and make corrections - help online users find data, and even adjust prices dynamically if it looks like the new price for raincoats still isn't getting customers to the racks. Smart systems also support personalization and customization. Your birthday, for example, could trigger all sorts of ads served up to you on your company's intranet or whenever you surf the Web. Smart applications can aggregate searches and then, based on rules you set, execute transactions. If you're looking for the best air fares, for example, a smart application – an **intelligent agent** – could aggregate data from all of the travel sites and then purchase the "best" one according to your definition of "best."

SFA and CRM front-office applications need automation, which sits at the back-end of the data organization, integration, and analysis process. Once patterns are discovered, once profiles are developed and once rules are specified, when specific conditions are met, then smart applications can, for example, automate a marketing campaign or follow-up a customer service inquiry.

Your infrastructure needs automation. Your computing and communications infrastructure needs to work together, reliably and securely, but all kinds of things can go wrong. What you want is insight into infrastructure operations with a set of rules designed to react to anticipated and unanticipated conditions. So when the amount of Web traffic is unexpectedly high, a rule fires that says: "when incoming sales traffic is high, and server capacity is low, then re-route or queue non-essential transactions." What does this rule imply? That you can monitor Web traffic for spikes, that you want sales inquiries to receive priority over other transactions, and that if unusually heavy traffic is detected, you want your customers to get through no matter what even if it means queuing large numbers of non-sales-related transactions. Ideally, your infrastructure knows about itself and how healthy it is. For example, it should know how many computers exist and what versions of which software are running on each one and which ones have too much and too little power. It should know which devices fail most often (and expensively or inexpensively) and take steps – alerts, for example – to anticipate failures. All of these kinds of back-office, front-office and infrastructure tasks can be at least quasi-automated. The larger argument I'm making is that the best management is "exceptions management," or the management of unanticipated events and conditions that you could not foresee.

The Internet was not developed to seduce people to live online. It's a communications and transaction processing channel that's fully capable of "self-execution" (which does not include suicide). Obviously not all transactions will be automated, but you should explore your ability to off-load processing to Web-based applications capable of interacting with employees, customers, suppliers and partners – and with your data about their preferences.

Our initial infatuation with the Web was more about consuming eye-candy than solving problems. Web site "look and feel" was the mantra of early sites, but over time discriminating surfers began asking questions about what the sites provided beyond "brochureware." Companies then got more serious about supporting transactions of one kind or another and that's basically where we are today: lots of sites with simple-to-complex transaction capabilities that are increasingly difficult to use. Some companies of course only live on the Web. Companies like eBay and Amazon are good examples of companies who live and die by the capabilities of their sites. But how much processing can be automated? On the business-to-consumer side, you could instruct Amazon to execute transactions on your behalf when certain things were true, just as a procurement officer could instruct a business-to-business site to automatically execute transactions with notification when the transaction is executed or if there's a problem with the transaction, such as the inability to get insurance or a problem with the carrier who's expected to transport merchandise.

The evolution of collaboration, supply chain management, personalization and customization will stimulate automation. As consumer, supplier, employee and partner profiles deepen, it will be possible – and desirable – to automate all sorts of customized transactions. Assuming that our privacy laws and preferences have been worked out, lots of us will empower retailers to execute transactions they know we'll like. It's the closest thing to a personal butler that many of us will experience, and the range of transactions – from the complex to the mundane – is theoretically limitless.

As I've already suggested, lots of automation is likely to occur through the assistance of hard working intelligent agents. Yes, I know, we've been talking about agents for years – almost as long as we've been talking about voice recognition – but we're so close now that you shouldn't allow your skepticism to push you behind the curve. Intelligent agents will come in several flavors. There will be really smart, powerful agents authorized to do all sorts of personal and professional things, and less aggressive agents that will only have basic transaction authority. Instead of debit cards, we'll give our kids mini-agents to help them plan their activities and manage their money. As they grow, they'll graduate to the smart, powerful agents that can help them manage their lives and plan their futures. No, I am not talking about deep intelligence here or the inevitability of digital friends or psychiatrists, but rather simple "if-then" production rule-based applications capable of – and authorized to – transact personal and professional business. Many of us are already part of the automation trend. Our bills are automatically paid each month and we get retail opportunities pushed at us from multiple sources, but we haven't authorized an agent to transact large chunks of our business. At work we automatically send and receive products every month but here too we're still manually executing the largest transactions.

Collaboration **requires** automation. Why? The sheer number of transactions that collaboration enables will require degrees of automated processing – unless we want to manually inspect every personal and professional transaction that occurs. While there are still arguments about just how pervasive agents will become, as business becomes more collaborative and the number of transactions grows, we'll need to rely on agents to manage these transactions.

While I realize that this sounds crazy, agents are likely to work directly for us but will also be members of larger agent societies. We'll tell them what to do and as members of the agent society they'll be able to perform the tasks we ask them to perform. As members in good standing they'll be admitted to wholesale and retail sites, banks, insurance companies and alternative forms of transportation. We'll finally have company as we work 24/7 to achieve whatever goals we think deserve this kind of commitment. This will be true on the business-to-business (B2B) and business-to-consumer (B2C) sides of transaction processing.

Our roles as transaction managers will be very different from our roles as transaction processors. Instead of entering data, we'll enter "rules" that will define what our agents can and cannot do. As they perform designated tasks we'll manage the exceptions, the transactions that go awry. All other transactions will process unnoticed.

How many of your company's transactions could be automated? Are the "rules" by which you conduct business well-defined and documented? Probably not, or probably in Charlie's head. Where's Charlie, and when's he likely to quit, retire or get whacked? Yes, there are other reasons to automate critical business processes.

One indicator of the validity of a business technology trend is what the companies that spend billions on research and development think about it. Actually, if they like a technology area, given the dollars they can put to work, they can create a trend. IBM and Microsoft are making major bets on what I'll call intelligent systems technology. IBM refers to the initiative as "autonomic computing," which is an inclusive concept that begins with mass data storage and ranges all the way to fully automated transaction processing. Microsoft has had intelligent systems on their target list for years. These initiatives have been blessed by the gurus – especially Tim Berners-Lee, the father of the Web and the guy who runs the influential World Wide Web Consortium (W3C) – under the term "semantic Web," which refers to the Web's next major capability: contextual intelligence. This kind of attention is non-trivial. When the people who spend the most money making things happen meet with the people who define what should happen, progress usually occurs. And there's another driver to track to determine if a business technology trend has legs – the military, which for years has invested huge amounts of taxpayers' dollars to make just about everything smarter.

What do you need to know about "automation"? First, it's a bona fide trend. Second, it's dependent upon other things like data integration, security and ubiquitous communications. Third, lots of people are spending lots of money to make devices, infrastructures and transactions smarter. Fourth, the application of intelligent systems technology is far-reaching from your infrastructure, back-office, front-office and e-business applications; and finally, collaboration requirements will fuel a lot of innovation in this area, innovation that you should track and pilot.

Trust

What a great topic. It turns out to be very real, very important and very much a differentiator. As the number of digital transactions grows, trust will become even more important primarily because of the physical distance between the transacting parties: if you trust that a meal will be good but it's not, you can immediately complain to the waiter and the chef, and probably get your food replaced or your money refunded. In this instance, trust doesn't need to extend too far. If it's violated you have instant recourse. But when you transact business at arms length, trust becomes a necessary and sufficient requirement for a successful transaction. When you buy something over the Web, for example, you need to trust the vendor more than when you buy something in person.

As transactions become personalized and customized, brands will assume new personal meaning. If I grant a vendor access to my PDA that vendor had better not abuse the privilege. In other words, as the number of communications channels with customers increases, so too does the requirement to efficiently manage customer trust.

Trust assumes security and security assumes the ability to authenticate users, protect data, control access to networks and applications, and of course the avoidance of annoying viruses. Trust is what you want your collaborative team to feel when they interact with you. If there's any doubt about your commitment to privacy or your ability to secure transactions – while protecting the infrastructure from viruses and other problems – your ability to collaborate will fail.

Business Convergence Scenarios

If all of the collaboration pieces are coming together how do you know when or how they will blend? All businesses are different and while the general forces of collaboration, personalization, customization and supply chain planning and management will influence your company, it's important to gauge the pace and direction of the collaborative change most likely to affect your world. Check out Figure 13. Where would you locate yourself? How do you get green? Do you have a strategic brain trust? What forms do alternative collaborative scenarios take in your industry? **Where should you be?** Where's the competition – and where's it going? How well do you know these things?

Figure 13. Collaborative positioning

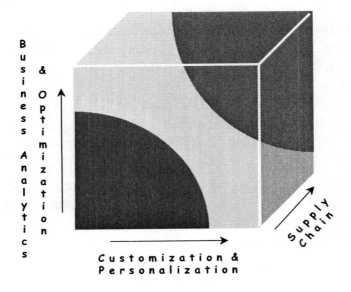

I have to tell you that after years of screwing around with "strategic planning" exercises, consultants and gurus, I've decided that "fast and cheap" scenario development is the way to gain insight into alternative collaborative futures.

Obviously, scenario-based planning can help reduce uncertainty about the future. The objective here should not be an elegant document suitable for publication in the *Harvard Business Review*, but a vehicle for provoking thought and then specific action.

So how do you build these things?

- You need to **monitor trends** that pertain to your industry on a regular basis, including social, political, workplace, work habit, regulation/deregulation, national and global economic, and macro-technology trends.

- You need to **monitor and report on known and anticipated competitors**. The good news here is that people love to watch the competition. This is because we benchmark and validate our behavior against our competitors. The huge danger is that if the competition you're tracking only extrapolates from its current business models - and that's all you track - you're likely to miss the out-of-the-box, unconventional competitors that can do serious damage to your market share. You should profile competitor business models and processes at least annually including details about the products and services they offer, their channel strategies, and their market niches. You also have to guard against assuming that the competition knows what it's doing. Some of them of course do, but some really don't.

- Based on all of the above – the environmental scan, competitor analyses, and existing business models – **future collaborative models and processes should be defined**. The key here is to simultaneously extrapolate from current trends **and** leapfrog current models into whole new collaborative business areas. How will you collaborate with customers, suppliers, employees and suppliers? What should your sales and marketing strategies look like? How will you cross-sell online and off-line? What role will automation play? How will you guarantee trust?

- **Key question** – do you have the expertise in-house to think creatively **and objectively** about collaborative business? If you do, beg, borrow or steal their time. If not, hire some vertical industry gurus (with deep industry knowledge - and empower them to speak candidly).

- After you've modeled the future, **prioritize** the collaborative processes, products and services that will keep you competitive. Keep the list short: a 20-item to-do list will always yield 15 orphans.
- Then **convert the prioritized processes, products and services into high-level collaborative business technology requirements** with specific reference to communications, applications, data, security, standards and people investments you'll need to make. You should also make a list of technologies, business processes and services that might be in-sourced, co-sourced or outsourced.
- **Create a "devil's advocate" process** where new collaborative business models and processes are reality-checked by nasty sons of bitches.

The objective of this process is to reduce - not eliminate - uncertainty about the future. The more uncertainty is reduced about the businesses you expect to be in, the more you can reduce investments in business technology. Without uncertainty reduction, you'll over-invest in business technology "infrastructure" (like desktop and laptop computers and networks) or completely miss the boat (how many companies failed to recognize the importance of TCP/IP – the Internet communications protocol?).

It's important to assign confidence to the "scenarios" of future products and services: likely products and services should be used as primary input to the business technology investment planning process, while those less likely should not drive the process.

The essence of business strategies is the confidence expressed in "bets" about future collaborative scenarios. Placing bets can be dangerous if they're based on faulty assumptions. At the same time, the right bets can save enormous amounts of money and lead to major competitive advantages. For example, assume that a national health care company targeted regional sales and service of managed Medicare based on the assumption that large regional differences among customers will require frequent product and service customization. This strategy would suggest a series of technology investments: distributed computing (where the customer databases were sprinkled throughout the regions), communications capabilities that would permit easy local access to customers and the ability to make changes to the systems that support enrollment, processing, and service – and of course a dramatically improved Web interface. A full commitment to regionalization and customization would **not** indicate additional investments in centralized mainframe-based applications that don't permit rapid customization. But you better be right. How many companies thought an ERP or CRM application would fix their problems?

Here's what the result should look like: several alternative scenarios within the collaborative space complete with a description of the drivers of each scenario, confidence levels and a detailed description of each scenario's dominant business model – all as suggested by Figure 14.

So what business are you in? What business do you do well? Which business processes will persist? What are the new collaborative business models that apply to you? Who's working on them? Will you lead or follow?

Let's step back and summarize a little. We've talked about collaboration, customization, personalization, supply chain planning, management and optimization, real-time analytics, automation, trust and how to better position your company within these trends. Scenarios can help you develop collaborative snapshots that will also help you better understand how and where business integrates with technology in order to facilitate continuous, automated transaction processing.

Collaboration will drive the need for technology integration, just as technology integration will enable collaboration. Just stay out of the red.

Figure 14. Alternative scenarios

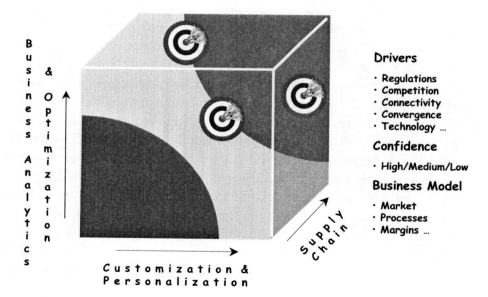

So What Does Everyone Think?

Feel free to think out loud. I know damn well that you C-level executives, directors and managers have all sorts of reactions to what we just discussed. Anyone?

The CEO ...
"Well, that's all well and good, this collaboration stuff, but it's not at all clear if you're right, where we are along some kind of collaboration continuum and what it will cost to get more collaborative ... got any proof?"

The Facilitator ...
"It's good to be skeptical because the biggest driver of collaborative change is efficiency ... this is not about 'cool' ... making supply chains more efficient saves money and increases margins ... customization and personalization increase sales ... analytics help you optimize investments, marketing, sales, service, production and distribution – and contribute to profitable growth by enabling the other activities ... all good stuff ..."

The CEO ...
"Still waiting for proof ... I could just sit here for a couple of decades doing exactly the same things we're doing now and make money ..."

The CFO ...
"You think so? I don't ... our cost structure is fragile – and our competition is aggressive ..."

The CMO ...
*"Can't sell the same message forever ... even if it **is** great ..."*

The Facilitator ...
"Look, we're talking about saving and making money ... we're not talking about making massive changes without rationale ... we're talking about forward thinking that assumes increasing collaboration among employ-

ees, customers, suppliers and partners ... and we're talking about the collaboration prerequisites ... look at this picture ... it tells us that concepts and commitment must accompany processes, procedures and technology ... that they must all exist for this to work ...

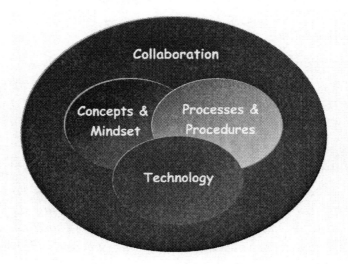

... that if you go out and buy supply chain technology, for example, without a clear implementation plan or without major buy-in from most of the companies in your supply chain, then you'll screw it up ..."

The Chief Operating Officer ...
"Huge implications here ... why not take baby steps?"

The CEO ...
"Makes sense ..."

The CFO ...
"Unless the competition moves faster and kicks our ass ..."

The CSO ...
"For the record, collaboration makes our security and privacy problems bigger ... you'll have to spend more money here to make our collaboration 'trusted' ..."

The CTO ...

"Very elegant architectures here ... I especially love the real-time analytical applications that enable optimization ... that ..."

The CEO ...

"Get back in your box, will you? This is not about 'elegant technology' ... who hired a 'CTO' anyway ..."

The Facilitator ...

"Here's another key picture ... look at the pieces and how they interconnect and ride on automation and trust ..."

The CFO ...

"Ever wonder why this guy's always showing us pictures? Doesn't he think we can read?"

The Facilitator ...

"Please ... I'm just trying to keep it simple ... if you want complicated hire a consultant ..."

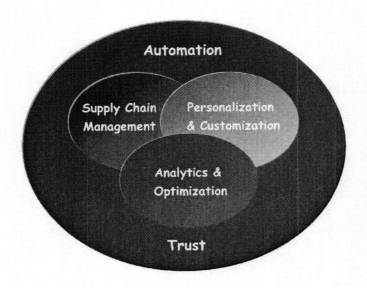

The CEO ...

"So the questions still are: 'where are we in this collaborative world, where do we think we need to be – and why'?"

The Facilitator ...

"Exactly ... do you know? ... do any of you?"

The CIO ...

"We have clues we can follow ..."

The COO ...

"We could build some scenarios ..."

The CFO ...

*"How much is **this** going to cost?"*

The Facilitator ...

"$100K for the process and several well-documented alternative scenarios ..."

The CFO ...

"Too much ..."

The Facilitator ...

"Really ... how much did you spend on McKinsey last year?"

The CMO ...

"Can you really automate this stuff?"

Endnotes

[1] The saga of the B2B exchanges is now well known. They were ahead of their time and failed to map the competitive vested interests that EDI-based B2B transactions had mapped for years. When smaller, independent exchanges appeared in the late 1990s the assumption was that they'd generate enough traction to win some significant market share. But when the entrenched buyers perceived the threat, they created their own exchanges and reminded their suppliers about who butters their bread. The net effect was the collapse of many independent exchanges (accelerated by the venture capitalists who stopped funding them as soon as the exchanges failed to generate meaningful revenue).

Chapter IV

The Technology Conversation – How The World Should Work

This is a monster chapter, with tons of stuff and just enough jargon to make you angry – even if you're a technologist. The challenge here is to focus on the important issues and avoid as many of the relatively unimportant ones as we can. Like anyone developing an agenda, this conversation has something for everyone but hopefully not too much of any one thing for anyone. Here's the list:

- **We spend too much money on business technology** – way too much. How come? Because we invest in the wrong stuff (and often do it stupidly). It's time to stop doing this and time to start investing with the uber (collaboration/integration) filters.
- **There are five not-so-easy pieces to this puzzle: applications, data, communications, infrastructure and security/privacy.** These pieces – along with the collaboration stuff we discussed in Chapter III - should occupy your time.
- Your **applications** must integrate (and you therefore must become real good at picking the right glue to make them all work together). There should not be hundreds or thousands of applications in your company; you should not adopt a "best-of-breed" applications strategy. You deploy applications that give you analytical insight into your collaboration envi-

ronment. You should follow the big vendors technology standards directions; and you should pilot some Web Services applications – the glue that just might make a major difference.

- **Data is still the lifeblood of your company, so** make sure it's clean, accessible and located in only a few "platforms." If you support Oracle, IBM, Microsoft and other large database platforms, you're insane, and you'll have to spend tons of money supporting these platforms as well as the data warehouses you'll have to build to be able to collaborate.
- **Communications is wired, wireless, narrowband, broadband and happens on desktops, laptops, PDAs, pagers and all sorts of converged devices.** Pay attention to the adoption of wired and wireless broadband, wireless collaborative applications and the integration of voice and data communications over IP (Internet Protocol) networks.
- **Security** has six not-so-easy pieces: policy, architecture, authentication, authorization, administration and recovery – and they're expensive. Get help here if you don't have the horses in-house to get the job done – **and worry about privacy** and its relationship with personalization and customization.

What Do You Know?

Hope you enjoyed the business discussion in Chapter III. We'll build on it here in some important ways. Hopefully, you'll start to see some major connections between collaborative business and integrated technology and therefore begin to embrace the whole business technology convergence philosophy (there's Kool Aid for the skeptical). The inter-connections between business and technology are inescapable. If you want to cross-sell, for example, you have to have a reliable relational database management system and data and applications integration technology. If you want to personalize and customize, you need lots of data technology and the inference models to "know" what individuals want. And if you want to enable your supply chain, you need a set of accepted standards and procurement software to make it all work. Again, make sure that the processes and discipline exist to get the most out of this stuff. If they don't, you'll become a Gartner statistic.

We're going to talk about the entire range of technologies in this session. Of course you probably already know about lots of them. No doubt you've paid for them year after year. How many of them left you uneasy? Isn't it amazing how often technology hype degrades to disenchantment? Let's look at a short list of over-hyped "killer apps": enterprise resource planning (ERP), network and systems management, e-business, sales force automation, and the most recent disappointment, customer relationship management (CRM).[1] Do you know what these applications have cost – and what impact they've had on your business? The reason why we talked about business trends first is because while the essence of business technology convergence is the intersection of two worlds, the *raison d'etre* for convergence is to stimulate profitable business growth. Questions about technology's contribution to business ride on answers to questions about business' clarity about technology's enabling role. While we've had problems, this stuff can work – especially if it's matched with the right collaborative business requirements.

Years ago I wrote about the "cocktail party phenomenon," or the CPP. The CPP, similar to the "flight magazine phenomenon" (or the FMP), is when senior executives discuss the latest technology trend while drinking the finest Chardonnay. After a couple of hours of talking big, they go home and sleep on what they've missed. How is it, they ask the next morning, that Charlie – who's pretty much of an idiot and runs a poor imitation of their wonderful company – knows all about ERP, CRM and jewels with names like Tivoli and Unicenter? And he's building a B2B exchange? "What the hell is that?" The senior executive is gunning for bear. "Why aren't we deploying CRM and why don't get an exchange going?" CIOs and CTOs dread these kinds of cocktail parties. (A similar series of events occur when senior executives fly somewhere and read in-flight magazines about how terrific object-oriented databases, biometrics, and streaming video really are – this is the dreaded FMP.)

OK, let's just admit it: most technology investment decisions are made on less than perfect information. More often than not, there are as many intangible variables as tangible ones. Keep in mind again that the Gartner Group reported that over 75% of all major software projects fail. So let's start with the grim financial context, let's get it all out. According to recent benchmarking research:

- On average, U.S. companies spend over $9,000 per year, per employee on computing and communications technology and support. On the high end there are companies that actually spend over $22,000 per year, per employee on technology. One study reported that some companies spend upwards of $38,000 per year, per employee on technology.[2]

- Between 1998 and 2000, companies overspent on high-end servers to the tune of $1 billion. The same study estimated that companies would waste another $2 billion from 2001 to 2003.[3]
- CIO Magazine reported that companies waste $78 billion a year on failed software projects.
- PricewaterhouseCoopers reported that over the past 25 years the number of failed technology projects that resulted in litigation has grown dramatically, with 48% resulting from warranty breaches, 13% from fraud, 11% from breach of contract, 9% from negligence and 7% from misrepresentation, among other problems.
- The National Institute of Standards and Technology (NIST) found that software bugs cost the U.S. economy almost $60 billion a year.

These numbers are intended to sober you up. Did it work?

The numbers are staggering. So how is it that we still invest over a trillion dollars a year in technology products and services when so much of it doesn't work? One answer is that investment criteria were relaxed or non-existent during the mid- and late-1990s. How many companies really scrubbed their e-business investments, for example? But as I've said repeatedly, even though lots of technology is mis-applied or out-and-out wasted, we're at a point where it's possible to not only avoid major mistakes, but to integrate technology and business in ways that were impossible five years ago. A classic conundrum. It's almost like trying to convince someone who's lost a ton in the market to buy a stock with a low P/E when it's at a five-year low. Most people shy away from it, even though they "know" that it will come back. Since 2000 we've been in an investment trough, where every technology investment had to yield a measurably huge return-on-investment (ROI) calculated on (low) empirical total-cost-of-ownership (TCO) data. We went a little too far with this requirement. There are still valid strategic reasons to deploy serious technology. But investments should always require due diligence (we'll talk more about "reasonable" ROI/TCO in Chapter VI).

Depending on your title, you see technology differently. Some of you see it as a giant sink hole; others as a way to differentiate your company from its competitors, your edge. Some see it as a sand box. Others see it as a necessary evil. What do you really know about it? About the structure of the technology industry? If you're a Chief Marketing Officer you should understand how software vendors "manage" versions to optimize their revenue streams. But if you're a CFO you may not fully understand how middleware works or why it's so important to your company, or how consultants identify problems that only they can solve. Are you surprised when it takes several years to install an application? Or when you hear about outsourcing lawsuits?

Some argue that technology is complicated especially when applied to ill-defined business models. I'd argue that our understanding of technology is both unfinished and unprofessional. We've been conditioned to think about technology as a silo – and we've managed it accordingly. Most companies still have "systems divisions" or "technology groups," when they should do whatever they can do to tear the silos down and rebuild integrated business technology organizations and processes (see Chapter V). We're evolving the relationship between business and technology. The whole point of the conversations here is to stimulate that relationship to the point where it's fully converged with collaborative business models.

We're unprofessional because we don't do nearly enough due diligence around technology investments, don't know how to measure ROI, and still make major technology decisions on the basis of incomplete and highly politicized information.

So let's not be constrained by the past and let's agree to try to start fresh. Let's also agree to keep the collaboration/integration filters front and center because they'll keep us focused.

Here's a way to think about all this that rounds out the collaboration/integration picture.

Applications Integration and Interoperability

There's a good chance that your applications portfolio is collaboration challenged. Remember the question about profitable transactions? The key is to identify your most profitable transactions and then dissect the relationship between the applications and those transactions.

Year 2000 compliance and e-business requirements drove enterprise applications strategy in the late 1990s. By 2000 companies were looking for ways to marry back-office with front and virtual (Internet) office applications. You still need to connect your employees with your customers and suppliers – and you'll

Figure 15. Collaboration and integration investment filters

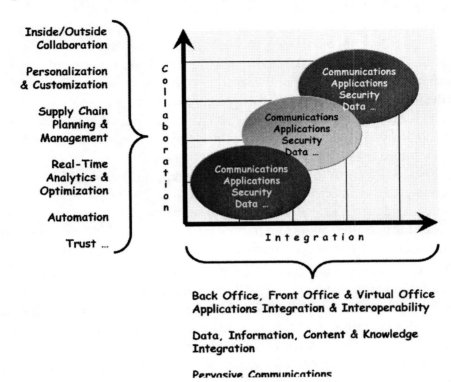

need to retain enough flexibility in your environment to accommodate enhancements and whole new strategies – as you juggle mobility requirements. E-business requirements should not drive your applications strategy. Collaborative requirements – of which e-business is a part – should drive strategy.

The applications end-game consists of a set of collaborative, integrated and interoperable back-office, front-office, virtual-office, desktop and personal digital assistant (PDA and other thin client) applications that support your collaborative business strategy. These applications should be standardized applications that support activities, processes, employees, customers, suppliers and partners regardless of where they physically sit or how mobile they are. All of your applications should be reviewed to determine their compatibility with this goal.

Like most companies who've deployed technology, however, yours probably did not initially or consistently worry about integration and interoperability.

You deployed accounting systems, claims processing systems, database management systems, word-processing systems, inventory control systems and shop floor manufacturing systems – one on top of the other. They maybe shared some data but did not work together. You could not take output data from one system and input it to another, or blast analyses in several directions for different systems to interpret. Real-time optimization was a vague concept and the idea that we could automate procurement, for example, was pretty bizarre. Consequently, we built these huge data centers that housed lots of island applications maintained by scores of talented (though sometimes humorless) technologists. But that was then, when business models were also disconnected, so it didn't matter all that much, until around 1975 when someone fired the first cross-selling shots across the bow of the USS Technology Independence. Yeah, that's right. Over a quarter of a century ago we began asking questions about integration. Fast-forward to the 21st century, and the questions – and answers – are no longer optional.

Look at your applications objectively. Which ones contribute measurably to collaboration through integration? Which ones directly or indirectly generate profit? Which ones require disproportionate support? It's essential that you assess your applications with reference to your collaborative business strategy and the relative contribution they're making to the company's collaborative business processes and profitability. If the outcome of that assessment is clear then decisions should be made to "decommission" (which means kill) applications (in the case of expensive applications that contribute little to the business) or transfer functionality to other, less-expensive-to-maintain applications (in the case of older systems with limited, but still valuable contributions to the business).

You need to assess the variation in your applications portfolio. How many architectures are you supporting? What's the distribution of functionality and architecture type? Do you have your most important applications on the oldest, most-expensive-to-maintain platforms? **You need a standard applications architecture.** If you keep buying and integrating different host-based, client-server, Internet, hybrid architectures your support costs will rise as rapidly as your reliability declines. But wait a minute. What the hell is an **application architecture**? Think about architectures as construction blueprints which detail the pieces and how they fit together. Applications architecture discipline is about the quality and consistency of the materials you use: if you use incompatible junk, the project will fail. This discipline applies to home, satellite and software applications construction. In practice, this means that your applications architecture has to be developed, approved, communicated and followed. In the trenches, they'll want to know if this means they should use an Oracle or IBM/DB2 database, BEA Systems' WebLogic applications server or IBM's, or commit to Java or Microsoft's .net. Yeah, I know. Here we go again. More techno-speak, more weird names. You don't need to know the innards of these tools. All you need to know is that a consistent, standards-based architecture will save you money, keep support requirements manageable, and provide the flexibility you'll need to become more collaborative. You need to think about the range of applications in your portfolio, how you procure, support, replace or modernize them. A framework that might help – and scare the hell out of you – appears in Figure 16.

Some companies – perhaps yours – have thousands of applications running on thousands of computers. Yes, thousands. Many of these applications have been around for decades. Some of them fly completely undetected below your technology radar screens. In other words, no one really knows who's using them, what it costs to support them, and why they haven't been killed. How does this happen? Here's how. Someone decides to move from one application to another, such as from the Lotus to Microsoft spreadsheet. Everyone "agrees" to switch over, but lots of Lotus diehards resist the change and keep using the old spreadsheet (and the Lotus applications they've created over the years). A Lotus underground develops. While there's no formal support for the old spreadsheet, there's a ton of informal help available to anyone at a moment's notice. What went wrong? Lots of things, probably including weak governance about applications upgrading, a lack of incentives to switch, poor standards setting and a general ignorance about the long-term costs of running multi-vendors/multi-versions of key applications. Happens all the time. Not only is this not-so-fictitious company running Lotus and Microsoft spreadsheets, but it's also running many versions of both.

Figure 17 presents the applications landscape with the collaboration/integration filters. Don't deploy an application if it doesn't support collaboration (supply chain management, customization, personalization, automation, real-time analysis, etc.) and migrate the ones with collaboration/integration potential to the green zone. Get rid of the others.

So what about all these expensive "enterprise" applications? Have you endured an enterprise resource planning (ERP) application implementation project? Have you paid for more user licenses than there are people on the planet? Have you experienced the joys of "shelfware"?

What about all of the new versions of operating systems and word processing applications? How often do you jump onto the upgrade treadmill?

Figure 16. The range of applications

	Corporate	Personal	
PDAs/ Thin Clients	• Email & Groupware • Calendaring • Browsers …	• Email & Calendaring • Personal Transactions • Instant Messaging …	
Desktop/ Laptop	• Word Processing • Presentation Graphics • Browsers …	• Financial Management • Communications • Knowledge Management	I n t e g r a t i o n
Enterprise	• Legacy • Packaged ERP • Internet, Intranet …	• Training & Education • Productivity Tools • Knowledge Management	
Management	• Network & Systems Management • Applications Management	• Project Management • Program Management • Workflow …	
Services	• End-to-End Services • Vorizontal Services • Hertical Services …	• Information Management • Searching • Configuration …	
Enabling Technologies	• Middleware • Artificial Intelligence • Components …	• Voice Recognition • Fingerprint Recognition • Artificial Intelligence …	

What about enterprise application integration (EAI) tools? What about portals and – the really big new question – what about Web services?

Let's back up a little and set the stage for what you need to do. First, some of your applications are bread-and-butter office applications, like Microsoft Word and Powerpoint, and Internet/Intranet browsers like Microsoft Explorer and Netscape Navigator. Microsoft Office is the de facto global standard for desktop/laptop personal productivity applications while browsers are essentially "open" which means that they work with pretty much any application that supports the hypertext mark-up language, otherwise knows as HTML. You also have big "enterprise" applications like SAP's R/3 ERP application or Siebel's CRM application. You also have home-grown applications that you've built over time and continue to support. And you have Internet applications that face outward to your customers, suppliers and partners. Finally, you have applications that help you manage other applications and your computing and communications infrastructure. Most of these applications sit on mainframes or servers and can be accessed from desktops, laptops, and

Figure 17. Applications investment filters

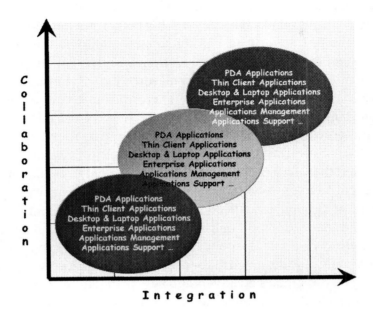

PDAs through wired and wireless networks. If you refer back to Figure 16 now, it might make a little more sense. (Figure 16 also identifies some of the enabling technologies that are fueling emerging next generation applications.) And they all have to integrate.

In order to enable collaborative business strategies you need to get all of these applications to work with each other, to integrate and interoperate. The conversation about strategic value versus operational support assumes collaboration and integration.

So what are the "issues" you should care about? Let's look at:

- Enterprise application integration (EAI) & migration
- Business intelligence & business analytics
- Portals & dashboards
- Key application technologies & technology standards
- Web Services – maybe the next "killer app"

And let's look at them all through the collaboration/integration lens.

Enterprise application integration (EAI) and migration is the multi-billion dollar business that you may never have heard about before. Essentially, it's the glue business – the business of getting databases and applications that were designed to ignore each other to become friends. This is very tricky, complex stuff and vendors and consultants blow all sorts of smoke around integration and migration problems and solutions. There are lots of companies that will come in and build custom glue for you that will connect a packaged application with a home-grown one, or two packaged or home-grown applications. If you go down this path you're certain to limit future flexibility – since the glue is "custom" and therefore only expandable by the vendor that created it (for another fee, of course). A better approach is to exploit some of the interfaces built into the more popular applications that make it easier to connect disparate databases and applications, though here too flexibility is dependent on how far the vendor goes to make their applications work with others. A third option is to deploy generic glue designed to connect lots of applications and databases, so-called agnostic glue. The advantage here is that it "works" with lots of databases and applications, but the reality is that it almost always needs to be tweaked to actually work in a production environment. There are vendors that specialize in generic glue – and glue application services.

The glue comes in many flavors. There's data glue otherwise known as data extraction, transformation and loading (ETL) technology. There's glue based on application programming interfaces (APIs). There's glue based on existing messaging applications often described as "middleware." And there's that generic glue known as EAI tools. What should you use? In case you're thinking that the computer industry is exceptionally screwed up, that the titans of the industry should have made the pieces work together, look at the automobile, appliance and aerospace industries. Do they have interchangeable parts? Are they standardized? Can you take a Ford engine and drop it into a Mercedes? We're not alone here. But the difference is – and this is significant – that Ford mechanics have no interest making their engines work in a Mercedes, and vice versa. We're stuck in computer and communications technology. We have to make the engines work together if we're going to achieve collaboration.

As you approach all this you need to make some major decisions about the number of older systems you want to keep and the investments you want to make in their ability to work with newer, packaged applications. Older, mainframe-based COBOL (one of the oldest computer programming languages) applications cost more than they make. One study reported that 60%

– 80% of technology budgets are spent on legacy applications and the mainframe systems that host them. Worse, all of the newer cost-effective technologies cannot support the older applications. The new "architectures" provide much more flexibility and scalability than the older ones. Bottom line: get rid of the damn things. Develop a migration strategy that gets you to the newer architectures and their capabilities as soon as possible. The collaboration business models we discussed in Chapter III assume integration and flexibility. Legacy applications don't integrate well and they're not nearly as flexibly (or scalable) as newer packaged applications. The best way to collaborate is to exploit the tools and techniques consistent with collaboration. If your CFO is opposed to migration it's because he or she assumes that the old systems can still do the job and that "patches" and "fixes" can keep the systems running for years, then you need to somehow quarantine these arguments so they don't infect your collaboration progress. If you want to collaborate, you'll need to deploy the newer technologies and stop investing through rear-view mirrors.

Enterprise application integration (EAI) and application migration are two of your new core competencies. This is not to say that you need to hire a gazillion EAI/migration professionals but you do need to understand a whole lot about what EAI is, how it works, what it costs and when – and when not - to deploy it. This stuff goes on your short list.

Business intelligence and business analytics are collaboration and technology goals. In order to achieve business intelligence you need to invest in business intelligence technology. Data integration makes this possible. The idea is simple enough. Integrate your applications to achieve collaboration optimized by business intelligence and business analytics, as suggested by Figure 18.

Business intelligence and analytics provide a window into your business and the means to adjust empirical insights into business processes and transactions. Think of business intelligence and analytics as the volume control on your business. But remember that in order to turn the dial you need to integrate your data and applications. These areas also go on your short list and should become core competencies.

What about **"portals"**? Where do they fit? Portals are at least three things: organizing devices, integration platforms and dashboards. If you have lots of applications – which you do – and crowded desktops and laptops that your users find difficult to navigate – which you do – and lots of applications that don't talk to each other – which you do – and the need to optimize the performance of these applications through business intelligence and analytics, and to see all of this in a dynamic dashboard – which would be great – then you might very well consider implementing a packaged portal as part of your collaborative business technology strategy.

Are you starting to feel like enterprise applications integration (EAI), migration, business intelligence, analytics and portal technology are all trending toward the same capabilities? Well, they are. And let's not forget the major back office vendors – like SAP – front office vendors – like Siebel – and virtual office vendors – like Ariba – that are also trying to make their applications more flexible and extensible through integration, intelligence and analytics. Many of these vendors offer their own portal products.

Advice? Stay focused on the integration/intelligence challenge. The means by which you achieve integration and business intelligence can vary. You can buy generic, agnostic glue, tools that convert data into insights, or portals that try to bring it all together in a single application which ultimately becomes an enterprise dashboard. How do you decide which way to go? Carefully and slowly. Shoot any in-house or out-house consultants that insist on an enterprise-wide deployment of any integration strategy. Pilot the alternatives and build empirical business cases around the results of the pilots. But regardless

Figure 18. Applications investment objectives

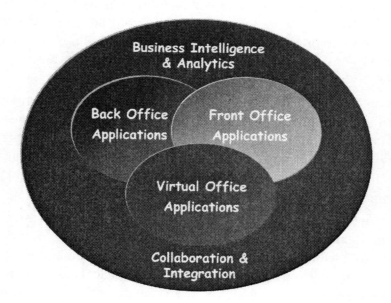

of which way you go, you'll need to develop a wide and deep understanding of all of the integration/intelligence technology that enables collaboration.

So what are the **key application technologies and standards** you need to track? With apologies right up front for the following techno-speak, you need to track the major application development and integration technologies offered by the major vendors and their henchmen, which include: XML (extensible mark-up language) and its extensions, Java (the generic programming language) and its extensions, and Microsoft's .net technology (designed to integrate data and applications). Tracking these macro trends will pay dividends downstream. Just remember that you don't want to be the earliest adopter of technology or what-appear-to-be-emerging technology standards. Why not? Can't you get an edge on the competition by going first? Very infrequently, and the risks of early adoption are significant, especially if you guess wrong and adopt what turns out to be the anti-standard! "Guess"? Yeah, that's right. Why in the world would you try to guess about which way the field's likely to move over the next few years? Step back and objectively track trends – but don't turn the tracking over to someone in your company who just loves Java – or hates Microsoft – because they do, they really, really do. Instead, follow the big vendors because they set the standards pace.

Let's talk about **Web Services**, which exhibits all of the characteristics of a trend that may or may not have long-term legs. The idea is simple: get the industry to adopt a set of common technology standards to make applications (and data) integrate and interoperate. Wouldn't that be nice? There are at least three XML-based standards that define Web Services: SOAP (Simple Object Access Protocol), WSDL (Web Services Description Language) and UDDI (Universal Description, Discovery and Integration). Sorry, here we go again. SOAP permits applications to talk to each other. WSDL is a kind of self-description of a process that allows other applications to use it. And UDDI is like the Yellow Pages where services can be listed. The simplest understanding of Web Services is a collection of capabilities that allow primarily newer applications to work with each other over the Internet. Because of the relative agreement about the standards that define Web Services, there's potential efficiency in their adoption. Conventional glue, for example, may consist of middleware, EAI technology and portals, where Web Services – because it's standards-based – can reduce the number of data and transaction hops by reducing the number of necessary protocols and interfaces. Eventually, the plan is to extend Web Services to your entire collaborative world, your suppliers, partners, customers and employees. As you may have already inferred, Web

Services – theoretically at least – reduces the need for conventional integration technology.

So what should you do about Web Services, what some call the industry's newest silver bullet?

You need to track progress, pilot some of the technology and wait for the industry to define the standards landscape. Do not commit massive amounts of capital to Web Services yet. Instead, identify some application integration metrics that will help you compare conventional integration approaches and technologies and those based on Web Services. Again, track what the big vendors are doing. Already IBM, Oracle, Microsoft and other vendors have announced their commitment to Web Services standards, though their actual commitments remain to be precisely defined. By now you also know that hype always precedes reality. Sometimes, a lot of the hype gets validated by actual capabilities, but just as often hype way outstrips reality. Do we need to remind ourselves about the dot.coms again?

Web Services is a fascinating technology development with – like so many others – enormous potential. But the reason why you need to track progress here so closely is because of the relationship between Web Services – standards-based integration – and collaborative business models. Web Services has real cornerstone potential. It could become a major enabling technology (or not). It's a classic high risk/high payoff trend. Let's just wait and see. Early returns on this technology are excellent; make sure you start working with Web Services immediately.

Data Integration

Data's still king but now we call it information, knowledge and content – and it's becoming dynamic. It lies at the heart of the new business models. Without data it's impossible to customize, personalize, up-sell, cross-sell, automate or gather business intelligence in real-time. But in order to achieve these capabilities, data, information, content and even "knowledge" all need to integrate.

Here are the pieces we need to understand. The collaboration/integration filters appear in Figure 19, while the elements of data alignment appear in Figure 20.

Figure 19. Data investment filters

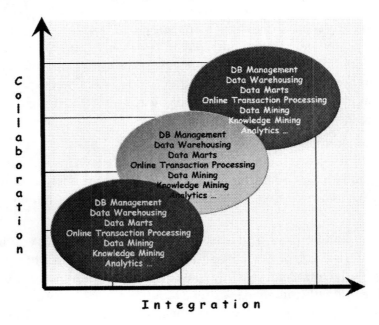

Figure 20. Data/information/knowledge landscape

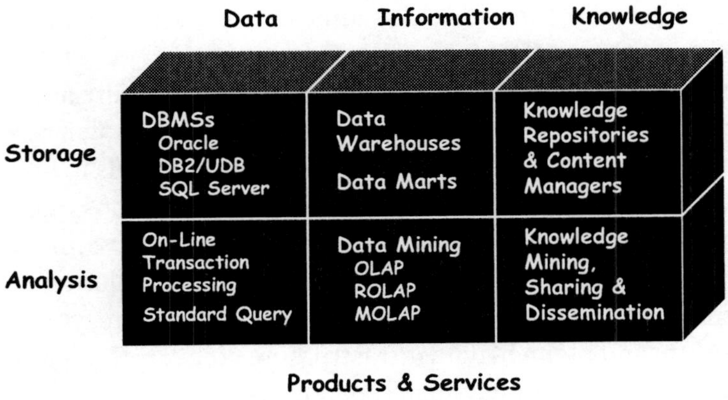

Let's look at data storage. Not long ago the list of viable database vendors included five or six major players, but now for all practical purposes there are three. In addition to the core DBMS applications are the hardware storage solutions you'll need to balance against expected data/information/knowledge processing requirements. Another major trend is the movement from hierarchical to relational database management systems and the migration from relational to object-oriented data base management. The more distributed your applications – and the greater your need for flexibility – the more you'll want to move to an object data architecture. Avoid supporting more than two database environments. If it's at all possible, select one. Only support three if you have stock in the companies because you're going to spend a lot of dumb money with these vendors.

Information storage options – data warehouses, data marts and special purpose hybrids – require some serious thinking about where you think you'll ultimately end up, how much money's available for the construction of these artifacts, what users will require and – relatedly – what the data mining tools will look like.

Knowledge storage is akin to dressing for a party to which you have no directions. Or – if you prefer – investing in a solution in search of a problem. The "knowledge management" business is just that, but the serious (read: measurable) pain it's intended to relieve is better described by the consultant doctors than by the victims (you). Nevertheless, you'll have to think about how to store unconventional, unstructured "knowledge" in order to play the knowledge management game (once you figure out what the problems are). Here's some food for thought. Rather than be flip about the young field, let's look at some of the assumptions. First, KM assumes there is knowledge to manage, that you've somehow codified the collective wisdom of your industry's and company's experiences. Second, it assumes that your culture and processes are sharing-centric, that is, capable of exploiting codified knowledge. Next, it assumes that you have – or are willing to invest in – the tools to make all this happen. Some vertical industries will be in better positions than others to exploit knowledge management. But some industries will have little or no need for what the consultants are assuring us is the next great revolution in database management technology. Look at your industry, your culture, your processes and your current and planned data infrastructure. If everything looks green, then do a KM pilot to validate your expectations.

On the other hand, there's lots of opportunity to exploit content management tools and applications. At some point, you'll need to move to a serious content management platform. But make sure your requirements justify the investment.

Storage is essential to analysis. But what's all this OLTP and OLAP stuff? Online **transaction** processing is what everyone's been doing for a long, long time. Online **analytical** processing – especially when coupled with data warehousing technology – is how data, information and knowledge get usefully exploited. It's the link with all of the discussions we've had about intelligence and business analytics. If you want windows and dashboards you need to invest in OLAP.

OLTP is the mother of all analysis. It provides insight into current internal data especially as it applies to operations and real-time transactions. OLAP provides insight when the focus is strategic, when the need is for reports and analyses, when access is unstructured and when optimization is an objective.

Information analysis requirements extend from OLAP's capabilities to desktop OLAP (DOLAP), relational OLAP (ROLAP) and multidimensional OLAP (MOLAP). DOLAP includes PC-based tools that support the analysis of data marts and warehouses. ROLAP includes server applications that support analyses cubed from a RDBMS or a data warehouse, while MOLAP exploits pre-developed data cubes.

Knowledge analysis and management is the end game of database management, data warehousing, OLAP and data mining. It's also at the heart of a learning organization. But it suffers from an identify crisis and should be pursued only when the criteria described in the Knowledge Storage section above are satisfied.

Data, information and knowledge infrastructure issues are complex because they're operational: all of this stuff has to work nearly all the time which means that you have to define and apply processes that can be implemented and maintained over time, and, of course, it all needs to work together.

A simple way to think about this appears in Figure 21. Everyone's got data in one place or another. Some of it's in an Oracle database, some in an IBM/DB2 database and some is still in Sybase databases. This "operational data" – especially if it's in different forms – often needs to get translated into a form where it can be used by any number of people in your company to perform all sorts of analyses. "Translation" results in the development of data warehouses and smaller data marts which support all varieties of online analysis and ultimately "data mining," the ability to ask all kinds of questions about your employees, customers, suppliers and partners.

So where's all this heading? Everyone's working on universal data access (UDA) from all tethered and un-tethered devices. Eventually, structured, unstructured, hierarchical, relational, object-oriented data, information and knowledge will be ubiquitously accessible. While we're a few years away from all this, it's helpful to understand the Holy Grail and to adapt your business models in the general direction of this capability. Microsoft, IBM and Oracle all have plans to provide UDA. It's important to stay abreast of their progress

Figure 21. Data → mining → customization, personalization, automation and real-time analytics

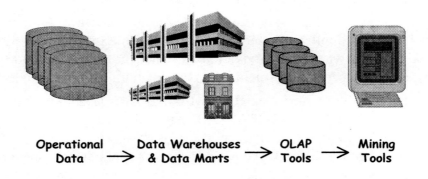

– and the implications to your business models and processes. Collaboration will require UDA and integration is the short-term path to that goal. Longer-term, if acquisition decisions are made properly, there should be less need to integrate lots of disparate databases.

Collaborative business models will drive data integration – whether you like it or not. You can't become collaborative unless your data (information, knowledge and content) is integrated. Over the years, you've probably deployed lots of different database management systems and lots of applications that had specific data requirements (for example, if you've deployed Oracle financial applications you have to run them on Oracle databases, which may or may not have been your preferred database platform – yes, that's a "gotcha!"). Consequently, depending on the amount of data variation in your company, you may be exquisitely ill-positioned for collaboration. Or, if you've had some discipline along the way and only have one or two database platforms, you're in a pretty good position to collaborate.

Your data integration efforts complement your application integration work. Some of the EAI tools include ETL, and vice versa. Investments in integration technologies should be driven by the results of your scenario planning exercises that position your company within the collaboration space. These scenarios will determine what applications you need and the extent to which the applications and data must be integrated. But regardless of where you find yourself in the collaboration space, you'll need to invest in data (and application) integration technologies.

Let's summarize. Your data must integrate if you're going to collaborate. If you have lots of different kinds of data in different places then you need to develop a data integration strategy which will probably involve building some kind of data warehouse. Once you build a warehouse you can conduct all kinds of analyses – analyses that facilitate collaboration. Over time, you need to reduce the need for all this integration by moving to fewer data platforms and standardizing the analysis tools – the tools you use to mine the data for collaborative insights and models.

Pervasive Communications

All of the emerging business trends we discussed in Chapter III assume ubiquitous, reliable, secure communications, and all of the technologies we're discussing here require integrated communications technologies. Collaboration and communication are almost synonymous. Personalization and customization require communications as do supply chain planning and management, real-time analytics and optimization, automation and trust. Some of this communication is organic and some is digital. The most effective is blended.

Let's look at the pieces.

As we move into the 21st century we need to recognize changing work models and processes such as telecommuting, mobile commuting, small office/home office (SOHO) computing, business-to-consumer (B2C) and business-to-business (B2B) transaction processing, internal workflow, groupware applications, business-to-employee (B2E) transactions, business-to-government (B2G) transactions, tethered and un-tethered phones, fax-based communications, electronic data interchange (EDI) communications, supply chain planning and management, continuous transaction processing, e-learning, customer service, supplier integration and partner management, local area/wide area/virtual private networks, the Internet, the World Wide Web and teleconferencing, just to name a few of the major challenges. A few?

We're changing fundamentally the way we work and live. Increasing numbers of us are working from home, from the road and in "virtual" spaces where we find ourselves constantly connected. New business models are also driving communications requirements, products and services. Companies are sending workers home, onto airplanes and into "office hotels" in an effort to reduce real estate and other support costs. The use of independent contractors is increasing, and more and more of us use communications technology to stay wired on a 24/7/365 basis. And let's not forget about the wireless tsunami.

Pressures on the businesses that require reliable, cost-effective communications - as well as the increasing number of individuals that rely on communications technology for personal information processing - are unprecedented and will continue to grow.

Let's start the communications technology discussion with a look at access. Right now you've probably got a hodgepodge of technologies that connect employees, some customers and fewer suppliers and partners than you'll need to connect in the near future. These technologies can be your friend or your enemy depending upon the clarity of your requirements and the infrastructure you're now supporting.

Access (to networks, applications, data, transactions) technology is central to your collaborative business models. Why? Because local and remote/tethered and un-tethered access will be necessary to support your employees working on the road and from home, your customers seeking information, service and – hopefully – transactions, and suppliers and partners who'll need access to your inventories and demand data.

Are you heading toward collaboration? Is it likely that your employees, customers, suppliers and partners will collaborate in the near future? You need to move toward a collaborative technology infrastructure capable of supporting anytime/anyplace communications.

But let's be clear about what all this is about. There's a distinction between shared communications and collaboration that's really important. For example, when you e-mail lots of people – and carbon copy even more – you're sharing communications, but when you create a "thread" of communication based on action/reaction, you're moving toward collaborative computing. In the near future, Internet bidding will become commonplace, triggering round after round of action/interaction. As that practice becomes widespread in your industry you'll need to be able to support asynchronous collaborative computing – not just shared communications.

The real questions have to do with the kind of collaborative environment you need to create and support, and the standards you'll need to adopt. While we have de facto groupware standards today, we don't yet have definitive supply chain standards that will make your total collaborative computing investments standards-proof. You need to pay close attention to external supply chain standards. Internally, there are lots of options – the key ones being LotusNotes and Microsoft Exchange – for supporting threaded discussions and workflow in your organizations. Lots of additional vendors provide workflow solutions.

The key here is decision-making that's holistic. Be careful about selecting workflow, e-mail, groupware or messaging technologies and products independently because they're all related. If you're hearing an argument for standardization here and an argument against best-of-breed products, you're listening. As our communications (applications and data) environments get more and more complex – and time to market becomes shorter and shorter – your time is better spent on optimizing applications rather than making a bunch of disparate products work together. (We'll talk more about all this later.)

Related to holistic thinking is the option of **unified communications** where all forms of communications occur via a single device. Eventually, it will be commonplace to receive faxes where you receive e-mail where you receive voicemail. Plan now for the infrastructure to support unified messaging. It might solve a lot of your communications integration and interoperability problems. What does this mean tactically? Stop buying personal digital assistants (PDAs) that aren't phones and pagers. The last thing you need to do is outfit your employers, suppliers and partners with chic leather belts on which they can hang three or four independent, unconnected devices.

All of this is useless unless there are **processes** in place to exploit your communications investments. Why so concerned about processes? Because without them investments are wasted. But processes are political. Prior to a new application, a new network and systems management tool, a new business model or a new communications network, new vested interests must replace old ones. Make sure that the processes necessary to support the new communications technology-driven business models are in place and make certain that the administrative and management processes to support them are well-defined and understood. (We're cruising quickly toward the business technology management discussions in Chapters V, VI and VII, discussions that will make a lot of this business technology stuff we're discussing here seem trivial.)

Assuming you know what you want to do and you've defined the processes necessary to make it all happen, you'll need to think top-down about the architecture you have today and the one to which you'll be migrating. Based on the collaborative business model that your scenarios tell you make sense, you need to move through this maze – migrating from the old to the new along the way – such that you end up with an integrated communications architecture. This means that none of the decisions you make can be made independently from any of the others. Your collaborative business model should guide you toward an access/connectivity and transaction processing communications strategy that converges. It must also guide you toward a strategy that supports migration from where you are today and leads to an infrastructure that works, can be maintained, secured and upgraded. Ask the tough questions and see who can answer them.

The movement toward virtual enterprises means that companies are forever changing their business models – how they sell, how they service and retain customers and how they measure success. This trend involves e-business but also sees the full supply chain kicking in within an adaptive communications network that will support local and remote customers, employees and suppliers. As you move more toward virtual/electronic business your communications requirements will grow dramatically.

The emerging collaborative environment we're discussing here creates opportunities and risks. The key is to anticipate your communications requirements within a larger trends context. Figure 22 puts it in perspective. The trick is to keep an eye on business collaboration and technology integration requirements as you make decisions about communications. It's that simple – or that challenging, depending on your people, your culture and your track record making these kinds of decisions (some companies almost always get them right, while others screw them up all the time ... and you are?).

Let's look at the implications of Figure 22. At the simplest level it represents a strategy, a high level filter that will point investments in one direction or another, but on another level it represents an opportunity to think about the drivers of collaboration and to ask some tough questions about where you're spending your communications dollars. For example, when you upgrade your communications infrastructure, can you be sure that the investments in bandwidth, network access and PDAs that you make will support supply chain planning and management, customization, personalization, up-selling, cross-selling, real-time analysis and automation in a trusted environment? When you add new communications technology layers, can you be sure that the layers will integrate and interoperate?

The approach to communications investing is the same as the approach to investing in applications, data, infrastructure and all business technology: it's got to converge with collaboration and it's got to integrate. If you're serious about these filters you'll save a ton of money and position yourself for a seriously competitive future.

Figure 22. Communications investment filters

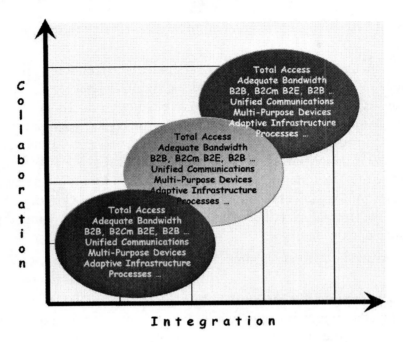

One more thing. Remember that business technology convergence assumes two-way influence. While emerging collaborative business models should influence our technology investments, emerging technologies also influence business models. Case in point? The aforementioned unified messaging. Sometimes technologies evolve under a "solution-in-search-of-a-problem" umbrella (or cloud) until the time that the technology begins to suggest some transaction possibilities. While we can certainly target collaboration, the particular flavors of collaboration (which play out as, for example, customization and personalization) can be defined by the capabilities of technologies like unified messaging.

Unified messaging is a technology that permits access, integration and interpretation of e-mail, voice and fax communications through multiple devices and access points. Imagine if your employees, customers, suppliers and partners could access all forms of communication from multiple devices and locations? How might you use that capability to provide customer service, answers to frequently asked questions, customized deals and cross-selling?

From a technology → transactions perspective, unified messaging permits all sorts of messages – including voicemail, faxes, live phone messages and e-mail

– to be retrieved from tethered and un-tethered phones (through technologies like "text-to-speech"), faxes, laptops, desktops, PDAs, the Web, e-mail and pagers, among other devices and applications. How might you use this capability?

The reason why executives need to understand broad technology trends and capabilities is to stimulate creative thinking about new business models and processes. We tend to think that creative business modeling comes from creative business modelers with wide and deep business experience, but lots of creativity comes from business technology convergence. But in order to exploit convergence one has to understand **both** forces well. With all due respect to CEOs, Presidents, COOs, CFOs, CMOs and some other chiefs, the depth of technology understanding is often unimpressive. Corporate boards tend to be even less informed about the direction or capability of computing and communications technology.

The recommendations here assume change:

- Make sure that the new collaborative business models translate into actionable communications requirements. Some of the requirements will be obvious, but others will be subtle, for example, like the ratio of in-house versus mobile users now and three years from now.
- Proceed holistically. Make sure that decisions about communications technology are linked. Tilt toward a standardized environment and away from a best-of-breed one: you don't have time to deal with endless integration and interoperability problems.
- Against a suite of requirements – like speed, availability, security, adaptability, and configurability – baseline your current communications performance and then project effectiveness against anticipated new requirements. Use the gap data to drive the new architecture, which will move you to (re-)consider wireless communications, fiber optic connectivity and gigabit Ethernet network connectivity, among other upgrades.
- Consider migrating toward unified messaging where all forms of communications occur via a single application and device. Eventually, it will be commonplace to receive faxes where you receive e-mail where you receive your voicemail. Plan now for the infrastructure to support unified messaging.
- It's also important to rethink your customer relationships. The anytime/anyplace possibilities of virtual connectivity must be assessed from whole

new collaborative vantage points. It's now possible, for example, to sell, service, sell again, service, cross-sell, service, up-sell, service – and then sell the data surrounding all of that activity to affinity marketers.

- Without the right processes, all is for naught. Well-defined, understood, communicated and approved processes will sustain your investments in communications technology. Without process buy-in, your technology investments will not pay dividends – and whoever is responsible for those investments will suffer politically.
- Your overall communications architecture – the access/technology and transaction processing technologies that you implement – must be integrated with itself, other technology and the collaborative business models and processes you plan to pursue.
- Institutionalize a process that reviews – at least a couple of times a year – the whole collaborative communications strategy (driven – ideally – by new collaborative business models).

Finally, there are a few emerging communications technology issues and trends you should track. They include:

- Broadband adoption
- Wireless technology & m-commerce
- The integration of voice & data communications

Our predictions about **broadband adoption** were wrong. Certainly the telecommunications vendors got it wrong. They incurred record debt to trench more fiber than anyone wanted. Where's broadband? Why isn't it everywhere – and why isn't it cheap? Broadband is one of those technologies that's in everyone's vested interest. But in spite of its lack of enemies broadband will only own around 35% of all connections by 2005. Watch this trend carefully: broadband adoption correlates with the adoption of collaborative business models. How so? Broadband facilitates wired and wireless supply chain management, personalization, customization and automation. Or put another way, if we had unlimited, cheap bandwidth how would it impact business? Yeah, tremendously.

Wireless technology, wireless technology standards, infrastructures and the whole world of "m-commerce" are redefining lots of our personal and professional processes and transactions. You need to know about this technology and what drives its adoption.

Recalling the discrete versus continuous transaction discussion in Chapter III, wireless communications will accelerate the movement toward continuous processing. Tethered and un-tethered connectivity keeps supply chains connected and facilitates mobile customization and personalization, business analytics and optimization. But there are limits to what wireless technology can accomplish. There are bandwidth and security problems: today it's hard to easily send and receive rich, multimedia content and there are major wireless security issues. So what has to happen for the situation to improve – for wireless applications to accelerate collaboration?

Welcome to the "Gs," which in the wireless world refers to the **generation** of transport technology that supports wireless communications. There's 1G, 2G, 2.5G, 3G and even 4G. In the early 1980s, analog cell phones used radio frequencies for voice communications. Remember the early clunky cell phones? They were part of the first generation (1G) of wireless technology. By the mid-1990s, analog yielded to digital encoding and we also got some limited text along with voice communication. This was the second generation (2G). Lots of people think we got stuck on 2G technology and there's lots of evidence to suggest that they're right. But just as we began to get really uncomfortable with 2G wireless technology, 2.5G technology began to emerge. 2.5G permits voice and text communications as well as access to the World Wide Web. 3G – 3rd generation technology – supports multimedia communications. It's available in a few countries and not in every area in those countries – including the U.S. 4th generation technology (4G) is so fast that it supports high resolution video. Unfortunately, 4G doesn't yet (really) exist. The Gs are distinguished by the standards that support them and the transmission speeds they provide. Why should you care about all this? Because the rollout of these generations affects what you can do collaboratively: speed and capacity define the texture of your wireless collaboration with suppliers, customers, partners and employees. Track this closely. When broadband wireless really kicks in, collaboration will be turbocharged. Track it and prepare now for its arrival. Ask yourselves what you'd do – today – with broadband wireless communications and then turn that list into a collaboration/integration project for someone to manage – first thing tomorrow. Among other business models, m-commerce would be on your list. Broadband wireless is one of those technology trends that you need to assume will arrive sooner than expected. If it doesn't, you'll be prepared. If it does, you won't be behind the m-commerce curve.

There's another trend that you should track closely: **the integration of voice and data on Internet Protocol (IP) networks**. This is one of those trends that enables multiple capabilities. If you could voice connect employees, customers, suppliers and partners through your data transport networks you'd save a ton of money and reduce your infrastructure costs. The addition of voice to your IP networks – Voice Over IP (VoIP) – will save you money and give you more flexibility than the current multi-network configuration we now use to collaborate, complex networks that carry voice, data, and applications that require management and support. While there are still lots of issues to resolve – like security, quality and inter-networking – the creative exploitation of integrated voice/data networks will provide significant competitive advantage.

Adaptive Infrastructures

Have you ever wondered what the hell technology people actually do all day? Well, lots of them take care of your company's infrastructure. You know, all of the PCs, the laptops, the PDAs, the networks, the communications, the Internet, the applications and the databases, among a few hundred other things. Remember that your computing and communications infrastructure has to support a wide range of collaborative business models and integrated technologies. How angry do you get when the network's down and you can't send or receive e-mail?

In order to make cost-effective infrastructure investments, several things should to be true:

- You have to know what you have in your infrastructure: the laptops, the desktops, the servers, the personal digital assistants (PDAs), the mini-computers, the mainframes, the communications network, the routers, the switches, the business applications, the messaging applications – all of it (and if you don't have an enterprise wide asset management system, now might be a good time to think about getting one).

- You have to know what the skill sets in your organization are for supporting and transforming your infrastructure.

- You have to know what plans are for collaborative applications, since chances are there's an "infrastructure gap" at your company.

- You have to know what it costs to run your infrastructure.

- You have to know what processes define your infrastructure support, how you acquire hardware and software, how you administer passwords, when you upgrade software, who works the help desk, and how applications are tested prior to deployment.
- And finally you have to know who pays the bills, what you have centrally-funded and what you expect your end-users to pay.

This is just a partial list, but you get the idea: in order to improve your infrastructure you have to baseline your current infrastructure assets and their performance.

But you also have to know what your infrastructure is expected to do. Are you primarily a back office, legacy applications shop maintaining aging applications in centralized data centers, or are you distributing your applications and making them accessible to remote customers, suppliers and employees? Depending on the answer, you'll need two very different infrastructures. Most likely, you'll need both: the former while you continue to support your legacy applications and the latter as you migrate to your inevitable collaborative future.

It's also likely that you have an infrastructure gap on your hands. Part of the reason is of course cost. No one wants to keep buying new stuff all the time and you're probably no exception: it's been easier for you to invest in new applications linked directly to business processes than "infrastructure" which everyone finds hard to define or appreciate.

Companies can influence the effectiveness of their infrastructures through the organizational decisions they make. Companies that separate infrastructure from applications often do so because they want the focus that segmentation creates. Unfortunately, in weakly governed organizations they also often set up conflict between those who build applications and those who support them. The reality? Make sure that your collaborative applications and infrastructures integrate and interoperate – and the planning for both are synchronized. (There are huge organizational issues here that we'll discuss in Chapter V.)

Many infrastructure managers think about infrastructure as computing and communications levels. Here are the three you should engineer:

- The **interface level** includes the desktops, laptops, browsers, PDAs and other devices that permit access to your data, applications, communications, messaging, workflow and groupware capabilities. You need to profile your current access "assets," including your desktops, laptops,

PDA and other devices used to access your applications and databases. You need to determine how skinny or fat your access devices need to be. You need to standardize on browsers and on an applications architecture that uses the browser as the common applications interface, that is, the primary way users (employees, suppliers and customers) access applications and databases through your communications networks. You need an asset management tool at the very least to identify what you have. You need to plan for an environment that will support an increasing number of skinnier clients and one that uses all computing devices as remote access devices.

- The **coordination level** includes the query, messaging, directory, security and privacy services that comprise your infrastructure. It also includes the transactions, applications and Web servers that permit you to support your applications portfolio. You may also invest in the tools and processes necessary to coordinate access and management. The most obvious tools are the network and systems management point solutions and frameworks – and make sure you track developments in Web Services. The goal should be to standardize on as few directory services, messaging systems, applications servers, and the like that make your applications work. Standardization can be vendor-specific or best of breed. I recommend you reject best-of-breed strategies in favor of more vendor-specific standardization. You'll need a network and systems management strategy which can be based on individual point solutions or on an integrated framework. Point solutions work best in smaller environments where network and systems management processes are hard to define and govern. Frameworks work best in large organizations where governance is strong enough to define and sustain processes. The implementation of network and systems management frameworks is complex and expensive. Be careful.

- The **operational level** includes the applications themselves as well as the applications management services necessary to keep transactions humming. It also includes the data/information/knowledge/content/metadata resources necessary to support transactions and applications. Your data centers reside in the resource layer of your infrastructure. But data centers should evolve to distributed data centers that (virtually) house distributed applications and data/information/knowledge/content as well as legacy applications and data bases that all must co-exist in the same infrastructure – at least until you kill off the mainframe/COBOL legacy applications.

Applications and communications architectures need to be designed and supported. Make sure that these designs are done holistically and with reference to the levels described above. Here's where collaborative business strategy requirements get converted into integrated technology designs that support profitable transactions. Integrated design is complex: make sure that you ask enough smart people to help design integrated communications and applications infrastructures.

Infrastructure involves hardware, software, and processes, and the discipline to faithfully convert strategic requirements into robust, reliable, secure and scalable networks, databases and applications. Take the long view here. Build an infrastructure that can adapt to evolving collaborative requirements. Do you know where your scenarios are?

It's difficult to find professionals who can support heterogeneous environments. Make sure that your in-house personnel are up to the task. If they're not, then consider outsourcing to a company that has the right mix of skills and experience. While there may be some reluctance to outsource your data centers, for example, remember that legacy data centers have been outsourced for years. As the number of enterprise and Internet applications rises, it's likely that legacy data center management processes will have to be substantially modified to integrate and support the newer collaborative applications (even as you plan to kill them off).

What works – and what doesn't? Just as it's essential to track your infrastructure assets it's also important to track their effectiveness. Without obsessing over return-on-investment (ROI) modeling, you should develop quantitative and qualitative effectiveness metrics that will permit objective performance assessments. Business cases should be developed prior to any significant infrastructure investments – and be prepared to pull the plug if the data looks bad.

What do you really need to know here? Infrastructure planning begins with a driving concept, a set of assumptions about what all the gear has to do. The argument I'm making – over and over again – is that strategic requirements are collaborative and tactical requirements are integrative. All decisions about infrastructure must converge with collaboration and integration. But what does this mean? Here's a short list:

- Access to your collaborative applications – supply chain, customization, personalization, business analytics and automated – needs to be ubiquitous, from any number of access devices (the access layer of your infrastructure). You need to architect the communications networks to work with all of your stakeholders.

- **The gear you buy has to support collaboration and integration.** Rethink buying anything that only does one thing, like a cell phone that's not a pager and PDA.

- **Rethink buying lots and lots of desktop computers (and certainly don't even think about replacing them for at least 36 months).** Laptops – while more expensive than desktops – make more sense for your collaborating employees.

- **This is important:** rethink the need for so much computational power on desktops and laptops. Does everyone really need the latest Pentium chip to send and receive e-mail? "Thinfrastructure" is a term you should explore. Years ago Larry Ellison talked about the network computer and how there was no reason why computational power could not be moved from PCs and laptops to servers on distributed networks. Well, he was right, if not ahead of his time. Begin to think about how you might migrate applications to network servers and redefine access and transaction processing around thinner devices (like smart, voice-enabled PDAs). So-called "thin clients" are cheaper and more reliable than Herculean desktops or laptops: while I cannot be certain about the general movement away from macho/macho machines, thinfrastructure logic is pretty damn compelling. I'd assign a couple of smart people to chase down the trends here and develop some scenarios about how thin clients might work in your company.

- **Take a look at the infrastructure management technology and applications on the market.** They can help you manage your applications, security, communications and even e-mail. These applications can function as monitors and managers and provide dashboard-like reports about how well your infrastructure is performing.

- **Make sure that your infrastructure encourages integration** and interoperability across your entire device, applications, data, communications and security environments. Make sure that whoever supports the infrastructure pushes collaborative processes and invests in integrative solutions.

Here's the deal. If you don't develop the right infrastructure you won't be able to deploy or support collaborative transactions. Not too many years ago – like the 1990s – companies actually delayed making their networks Internet Protocol (IP) compliant. I have no idea what these companies were thinking. Maybe – like the senior executive I mentioned in Chapter II – they actually believed that the Internet was a fad that would go away. Who knows? But here we are again. There are macro trends in wireless technology, thin client computing and infrastructure management that you need to track – and embrace. Separate the cool trends from the trends that will accelerate collaboration.

Security and Privacy

There are a set of technologies, processes and services that together constitute your security strategy. Related to all this is your privacy policy, which you better take seriously. Why do I say this? Because many companies are publishing privacy policies because the government tells them to, but they're not protecting the data of their collaborators as carefully as they should. There's a natural tension between personalization and privacy. As we discussed in Chapter III, there are real issues here, and we all need to make sure that we get the appropriate permissions to build our mass customization strategies.

Figure 23 maps the security landscape. There are lots of cells in this matrix.

Every one of them demands your attention because they all need to work together.

Let's look at the cells.

First, there's **policy**. Here are the guidelines:

- You have to write it down. All of your security and privacy policies and procedures must be codified, communicated and updated on a regular basis.
- Your security policy shouldn't be a bible, but it needs to be specific enough to reduce ambiguity. Rather than spend two years writing the perfect security or privacy policy – one that covers every aspect of your security environment – spend a lot less time and get one out that works – and make sure that you offer training around it.
- If you can't write a credible policy in-house, outsource it.

Figure 23. Integrated security requirements

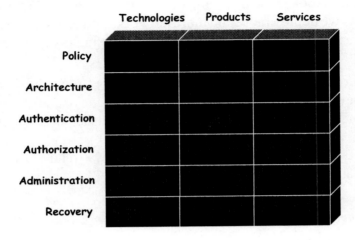

- The document should include – at a minimum – policies that address:
  - Data access
  - Applications access
  - Network access
  - Software
  - Privacy
  - Business resumption planning
  - Systems design & development
  - Risk assessments

Authentication is absolutely key:
- Passwords work, but can be expensive to administer and complicated for users to use on a regular basis, especially if they must remember multiple passwords.
- If your password policies are well developed (for example, requiring users to change them every 30 days or so, or the systems automatically deny them access) then they might work for you for the next few years. But if they're complicated and cumbersome to manage, you might consider alternative authentication methods, tools and products like smart cards and biometric devices.

- Single-sign-on to networks and applications remains a worthwhile objective. Take a look at the available tools and the processes that support them. As we deploy more and more networks and users traverse more and more networks and applications, you'll need single-sign-on capabilities.
- Make sure you investigate the range of available firewall technologies since they're changing all the time. Note the trend to embed more and more functionality into firewalls, functionality that includes lots of flexible authentication (and authorization) techniques. Over time, you should be able to off-load lots of functionality onto your firewall (and other hardware devices and software applications).

Who gets to do what? **Authorization** follows authentication:

- Once users are authenticated, they need to be monitored according to some pre-defined authorization schema. Access to networks, applications and databases needs to be defined and individual and classes of users need to know where they can go and what they can do once they get there.

All of this stuff is great until you have to support it. All security (and privacy) policy, authentication, authorization and recovery requires **administration**:

- Make sure that you ask questions about administration each time you consider a method, tool, technique or process.
- Develop some metrics against which you can track the effectiveness of your administrative procedures. Track the data over time to determine the cost-effectiveness of whatever administrative processes you put in place.
- Some basic administrative reports include:
  - User Sign-On Error Reports
  - User Policy Violation Reports
  - Resource Activity Reports
  - User Access Reports

Recovery is as essential to security as authentication. Make sure you don't short-change your security strategy by under-cutting investments in systems recovery and business resumption planning. Business disruption and resumption planning simulations should be conducted on a regular basis – at least twice a year – to determine if your business resumption planning policies and procedures will actually work when a major business disruption occurs. The basic elements of a business resumption plan should include:

- Plan Activation Policies and Procedures
- Individual, Group and Team Recovery Policies and Procedures
- On-Site/Off-Site Resumption Policies and Procedures
- Administrative Policies, Procedures and Responsibilities
- Contingency Planning

Supporting your security and privacy policies and technologies is tricky. Unless you have a lot of in-house security talent, you might have to look outside for end-to-end security solutions integration. This decision must be made carefully, since there's a tendency to think that the in-house staff – who may have managed security in a host-based, data center enclosed environment pretty well – can manage a growing number of distributed applications that link employees, partners, customers and suppliers.

The range of necessary skills is broad and deep. You'll need to make sure you cover all of the bases – and well. The key is the integration of the services into an adaptive solution. If your security strategy consists of lots of elegant pieces that don't fit well with one another you don't have a viable security strategy.

What else do you need to know about security and privacy? You and your team need to understand:

- Firewall technology
- Anti-virus technology
- Certificate authority technology
- Biometric technology
- Encryption technology
- Privacy compliance technology

These technologies enable your security strategy. You need to assign resources to track developments in the technologies and the products they support. While there might have been some skepticism five years ago about declaring distributed security a core competency, today the opposite argument should raise eyebrows. It's imperative that you understand the technology and how it interacts with your business model, applications portfolio, communications architecture and other dimensions of your infrastructure.

Privacy will become increasingly important, increasingly the focus of government regulations and – therefore – the object of computing and communications standards. Pay close attention to these trends, since the flip side of business security is personal privacy.

Change is inevitable here. Make someone accountable for developing security scenarios two-to-five years out.

The threat of information warfare must also be taken seriously. If you screw up your security architecture and infrastructure, your competitors will find ways into your networks, your applications and your databases. Is it easier to spend years working to increase market share or spend weeks destroying the competition's databases? Information warfare is a real threat. You should invest as much as necessary to protect your business – and in business recovery processes and technology should your defense break down.

You need to let go of the notion that security is a "step" you take (when designing, developing and deploying networks and applications), or that security can be serviced by lots of tools and techniques surrounded by distinct processes. While these notions are theoretically correct, they miss the long-term point: security is not a part of a network or an application, it's embedded in networks and applications. In other words, security is as much a part of a network as a router or switch, or as much a part of an application as a user interface or database. When you stop looking at security as a disembodied part of your network and applications infrastructure – but rather as an integral ingredient in an otherwise pretty complex soup – then you've achieved the next level of distributed applications development and management.

Yeah, another core competency to add to your list.

Figure 24 summarizes the security challenge.

Figure 24. Security investment filters

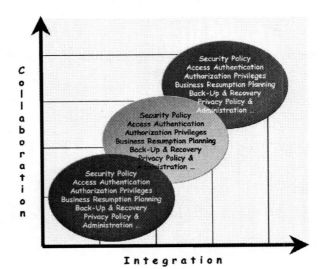

Business Technology Convergence

The point of this long discussion? There's a ton of technology out there and it all needs to work together to support your collaborative business models. Simple enough, huh? Unfortunately, the range of applications, data, communications, infrastructure and security technology is enormous – and growing. So how do you organize it all?

The collaboration/integration filters are really your friends here. If you accept the trajectory of your collaborative business models and the need to integrate disparate technology (and avoid deploying technology that doesn't cooperate), then the selection criteria are pretty clear.

Opinions?

The CEO ...

"I love drinking from fire hoses ... are you kidding? Way too much to think about – to worry about ..."

The CFO ...

"And way too much to pay for ..."

The Facilitator ...

"You're already paying through the nose ... you just didn't know how bad it was ..."

The COO ...

"All I know is that I can't do business when the technology's down ..."

The CIO ...

"And like it's always down? Give me a break. If we funded this stuff adequately we'd have more reliability ..."

The CSO ...

"And security ..."

The CEO ...

"How many database 'platforms' do we have in the company?"

The CTO ...

"Five ..."

The CEO ...

"Why do we have five and not two ... and how many applications do we have ... and do they talk to each other ... and do we have a portal and if we do, why do we have a god damned portal ... and what are we doing about wireless communications ... and broadband and voice-over-IP – whatever the hell that is – and security ... and privacy ... and what about the 'layers' of our infrastructure ... and ... ???"

The General Counsel ...

"Is there a point to all these questions?"

The CFO ...

"There sure is: if we're going to compete we have to optimize our business technology investments ... we're clueless here ... we have god knows how many data base platforms – whatever they are – applications, networks, desktops, laptops, PDAs or whatever ... what's the plan, what's the strategy? Standards? Single-source? Metrics? These are all no brainers!"

The CTO ...

"Well, not everyone would agree ... there are all kinds of debates about this stuff ... it's not that simple ..."

The CEO ...

"You sound like a vendor – or a consultant – telling me that I don't know enough to know how much help I need ... this is really starting to annoy me ... I'm starting to feel that we've left a ton of cash on the table over the years, that we can do a whole lot better ..."

The Facilitator ...

"Amen ..."

Endnotes

[1] There's a lot of "evidence" about these "disappointments." The Gartner Group publishes lots of total-cost-of-ownership (TCO) and return-on-

investment (ROI) analyses (www.gartner.com), as does The Standish Group (www.standishgroup.com) which both generally report that we often pay more than we should for hardware, software and services. The **National Institute of Standards & Technology** (www.nist.gov) published a report that software bugs cost users and vendors almost $60B annually. The Nestle vs. SAP case is widely known (see Ben Worthen's Nestle's ERP Odyssey, *CIO Magazine*, May 15, 2002), as are other failed enterprise projects (see Kim Girard's report on the Department 56 vs. Arthur Andersen - "Blame Game," *Baseline Magazine*, March 2002). Litigation sometimes results from implementation problems. See PricewaterhouseCoopers' "Patterns in IT Litigation: Systems Failure (1976–2000)." Also see Meredith Levinson's, "Let's Stop Wasting $78 Billion a Year," *CIO Magazine*, October 15, 2001, and Charles C. Mann's, "Why Software is So Bad," *Technology Review*, July/August 2002. Paul Strassmann's work is again relevant here. See www.strassmann.com for tons of insight and data. While there's are any number of horror stories out there, there are also some huge success stories, not to mention the everyday success of word processors, presentation packages, databases and e-mail. Finally, regarding CRM, see Michelle Schneider's, "CRM: What It's Worth," *The Net Economy*, February 5, 2001, and Rich Cirillo and Dana Silverstein's, "Can CRM Be Saved?," *VARBusiness*, February 4, 2002.

[2] See "The Book of Numbers," Hackett Benchmarking | Solutions, 2000 (www.answerthink.com/hackett for more details).

[3] See www.gartner.com for the report on server over-spending.

Chapter V

The Turf Conversation- Who Does What To Whom

Not too many controversial topics here. Here are the turf topics we'll discuss:

- Figure out how to **neutralize the politics around turf**: I have no magic here … it's been going on since people began to congregate … but take it seriously because it undermines effectiveness.
- **Forget titles**: you have to organize your companies to collaborate, support collaboration and enable it with integrated technology … you may or may not need "chiefs," or "directors."
- **Forget about consensus-based decision-making** in flat management structures … forget about big teams.
- **Command and control works** … even in decentralized organizations – which, by the way, I'm not that crazy about.
- **Innovation is special**: make sure your organizational structures encourage business and technology innovation.

Watch Your Flanks

Batten down the hatches. We're going to talk about organizational structure. Isn't it interesting how everyone perks up when these conversations begin?

How many management revolutions have we endured? How many of us still have a completely unacceptable sane:jerk ratio in our companies? How many times have we re-organized when a new management team arrived? Do we learn from our organizational mistakes? Have our management structures changed with the times?

This last question is key. If you're feeling more comfortable with the whole collaboration and integration discussion – the whole business technology convergence argument – then it may be time to determine the extent to which your organization supports collaboration and integration, along with the traditional functions of your company.

From another perspective, how many of us wrestle with the "who-should-report-to-whom" question several times a year? Seriously, have your efforts to "re-organize" the business technology relationship been proactive or reactive? Often because some influential people complain about the relationship, things change. But reactive changes usually don't last long.

Other changes are triggered by real problems with reliability or security. Context changes everything. While we all recognize the importance of terrorist threats today, why don't we recognize the potential of collaboration? And the necessary integrated technology to enable it? It's always about the big questions – but we have to see them before we can answer them.

There are also lots and lots of political agendas out there. We've all learned to be very careful about friends and enemies and the times when it's impossible to distinguish between them. Not so many years ago business people wanted their technology strong and cheap and technologists wanted the respect, time and money they deserved. Uneasy partnership? Absolutely. But the convergence argument I'm making here says that not only are the two in the same camp, but they each have to get wood to keep the fire going. They're teammates. The vested interests should be defined around this partnership, not around obsolete adversarial relationships.

Yeah, I know. Naïve. But is it? Adversarial proceedings are nearly impossible to manage. Maybe it's time to converge vested interests rather than play them off against each other.

You Report To Who?

Here are a few of the organizational issues worth discussing:

- **The re-engineering of business technology organizations** will surface as one of the major corporate imperatives of the new millennium. Companies will look to technology to (really) integrate with the business and provide competitive advantage through efficient collaboration.
- "Good 'ol boy" (and gal) **relationships will be (partially, never completely) replaced by strategic partnerships** that will be judged by performance – not historical inertia.
- As skill sets become obsolete faster and faster, there will be **pressure to change organizations at a moment's notice**. This will dictate against large permanent in-house staff organized to protect their existence. New applications pressures will kill entrenched bureaucracies and give rise to a new class of results-oriented hired guns.
- Companies will find it increasingly difficult – if not impossible – to keep their staff current in the newest business technologies. This means that **companies by default will have to outsource certain skills.** The approach that may make the most sense is one that recognizes that future core competencies will not consist of in-house implementation expertise but expertise that can abstract, synthesize, integrate, design, plan and manage.

How many of the above drivers hit home?

If you live in a decentralized organization – where the central technology organization owns the enterprise computing and communications infrastructure and the lines of business own their applications – pay very special attention to organization structure. Unless you're prepared to fight lots of religious wars between central technology and the lines of business, organize your internal technology professionals in ways that support conventional and strategic business models and processes. If your current organizational structure in any way, shape or form encourages an adversarial relationship between central technology and the lines of business then your organizational structure is flawed. The hell with "flawed": your organization will lose money and under-perform.

You have a number of organizational options. As your collaborative business evolves, it's essential that you undertake a brutally candid assessment of your core competencies today and – especially – what they should be tomorrow, and then begin to define the organizational structures that will exploit the right competencies.

You should also look at the centralization/decentralization, standards and governance issues, as well as architecture, infrastructure and metrics (in addition to return-on-investment and total-cost-of-ownership data). These **are** important issues and the challenge remains to organize your company in ways that result in more natural resolution of thorny issues. Is this a general management theory? Probably not, but it's actually fairly profound: if you've organized things so that big, important problems get solved with a minimum of effort then you've probably done something pretty good (and the converse is insane – where the smallest, stupidest problems get all of the attention).

The way companies try to fix their technology organizations and the relationship between technology and business is to keep adding pieces to what they think is a solution without realizing that incremental, sequential approaches to problem-solving are by definition unable to solve any problems (unless you're trying to fix a lawnmower that won't start). Yes, this is a set up for a more holistic approach to organization, and not just business technology organization but the organization of the whole company.

Organize – Because We Have To …

The assumptions we've made over the years about organizing technology to support the business were fine for a couple of decades, but then began to break down. The erosion took the form of arguments about the need for business/technology "alignment." Magazines still publish articles about the need for alignment. Hell, we have entire conferences about alignment in exotic places that attract hundreds of people (I think I answered my own question). There are even cruises to no where for technology executives and vendors who talk endlessly about how to make technology work better for business. But by and large, all of the alignment models are separatist models.

Without any sense of how radical you think these conversations have been, let's ask some potentially radical questions:

- Do you need a "CIO" or a "CTO"?
- Are your technology leadership positions well-defined?
- What do the business managers think about technology at your company?
- Do you rotate your business and technology professionals?
- How do you process collaboration requirements?
- Who owns technology integration?
- Do you know what you're good at – and what you do poorly?
- Are you strongly or weakly governed?

Enough of these drills. Organizational strategies should be holistic and as consistent as possible with business and technology trends. At a general level, we know that collaboration is a trend. We also know that collaboration is driving integrated technology, just as integrated technology is driving collaboration. Need another example? We want to use the Web for customer service. It's cheaper and it's available 24/7/365. The extended business model makes sense. But technology goes the process one better, adding co-browsing – the ability for a customer to synchronously interact with a customer service representative – to the Web site to help site visitors get out of trouble when searching for something or filling out a form. Very few business professionals are current in co-browsing technology. But when the technologist understood the requirements, an immediate leap was made to co-browsing technology that could enhance the online customer experience.

So what do you need to do? Figure 25 presents some ideas – and suggests how you might want to organize your company (presumptuous, huh?).

The conversations here have been about collaboration and integration, not explicitly about sales, marketing, finance, manufacturing or distribution, but there's no way I can talk about organization without including everybody.

Figure 25 suggests that you consider three pieces to the puzzle – formally. While we all understand traditional corporate functions such as sales and marketing, not all of us appreciate how fundamentally different the collaboration and integration challenges are. The suggestion here is to organize around the three (collaborate/support/enable) areas. Just three. This means that all of your business models and processes will be redefined along a collaborative continuum and managed accordingly. It means that support for these models – finance, sales, marketing, etc. – will be organized with reference to the

Figure 25. Organizational requirements

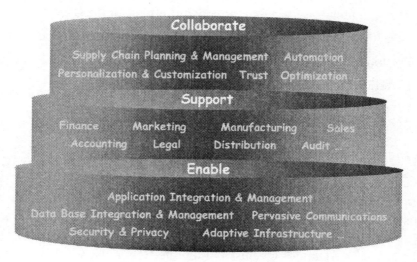

collaborative business models they serve as well as the integrated technology that enables collaboration.

In addition, the organizational model assumes an integrated collaborative, support, enabling technology strategy, that no piece acts independently – or at least not independently with impunity. Best practices round out the picture.

So what does this mean? Am I suggesting some gonzo organizational structure? Is this a not-so-subtle plot to get you to overthrow your current organizational blueprint? Not at all. In fact, Figure 27 suggests how you might convert the functional requirements into a workable organizational structure that's really not all that radical – though it would trigger massive changes in how responsibility and authority is allocated in your company.

I know, I know. It looks weird. Let's discuss it.

The middle – support – activity is what we're all familiar with. Here's where sales, marketing, manufacturing, distribution, human resources and legal live. Someone needs to run this organization and since I've already complained about how many chiefs there are, let's call this person the Support Officer. So far, so good. Then there's the Collaboration Officer, who runs the collaborative business strategy, models and processes. This is where supply chain planning and management, customization, personalization, automation and

Figure 26. Organizational requirements

Figure 27. Proposed organizational structure

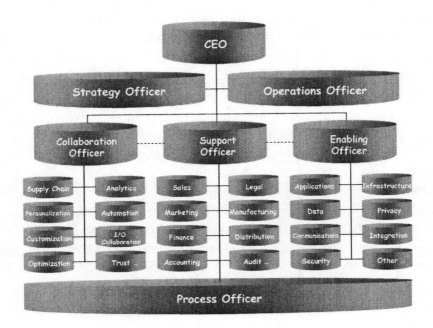

optimization live. Existing business models and new collaborative models would belong to the Collaboration Officer.

The Enabling Officer is what we would have called the CIO, CTO, the Chief Applications Architect, the chief (what else?) of technology operations and anyone else connected with technology if we could have morphed them into a single officer. Yes, this means that the "CIO" and "CTO" as we know and love them today get new titles and consummate their relationship with the business in a new organization that will have killer integration capabilities where "integration" is understood to mean integration of disparate technologies **and** integration with collaborative and supporting business models and processes. The EO also owns the applications integration, pervasive communications, data integration, technology infrastructure management and other things technical. The EO does not, however, exclusively own technology acquisition (outsourcing), just as the CO and SO do not own business process and support acquisition (outsourcing) decisions. Acquisition is controlled by the PO (with input, of course, from the other organizations).

We've also got another officer lurking around. The Process Officer has broad "best practices" responsibility – **and authority**. The Process Organization owns lots of things: project management (in which a Project Management Office can be housed), standards, research & development (and therefore innovation), measurement and metrics, including total-cost-of-ownership (TCO) and return-on-investment (ROI) data collection, outsourcing, business case development and dissemination, and all of the other appropriately enterprise or "neutral" activities that should not be located directly in collaboration, support or enabling technology.

This structure attempts to organize things in ways likely to keep the number of religious wars to a minimum: it keeps most of the foxes out of the henhouses, or at least it tries. The Process Officer – let's make this one a Chief – sits in the middle of lots of enterprise activities. But note the processes do not distinguish among collaboration, support or enabling missions. They're all treated equally. Outsourcing, for example, could be business process outsourcing or the outsourcing of desktop computer support. The perspective of the CPO is that outsourcing must integrate across corporate missions. Another frequent war – standardization – is also fought – with nuclear weapons – by the CPO. If we're anywhere in the evolution of business technology logic and sanity, we're all about standardization, as well as related migration policies.

What about the Strategy Officer and the Operations Officer at the top of the chart? What the hell do they do? The SO is a senior staff that makes sure that:

(a) there is a strategy and (b) the collaboration, support and enabling organizations stay consistent with it. The OO is the execution bully who makes sure that operations run smoothly. The CEO thus has five direct reports, but the SO(1) and OO live on the same block as the CEO, not down the street with the CO, SO(2) or the EO. You get the picture.

Also note that this structure is self-contained. The idea is that a line of business should have its own collaborative strategy, its own support, and its own integrated enabling technology. The officers of each of these activities are on an equal level. This means that debates about who technology should report to are pointless: technology, like traditional support, and the business transactions they serve, all report to the CEO – they all have big seats at the big table. The CEO protects the integrity of the organization by managing effectiveness through an enterprise strategy and an enterprise operations staff, and everyone's beholding to a (seriously) powerful process officer. The enterprise strategy, operations and process officers are equally powerful and together keep the collaboration/support/enabling technology activities "focused." In a single or double line of business model the self-contained model can work. When there are multiple lines of business it gets more complicated (see the following for a couple of cute pictures on this).

I also have some – well, strong – ideas about culture and power – the real organizational drivers. I'd argue that "empowerment" has yielded mixed results. While it makes sense to build consensus whenever possible, it also makes sense to move decisively. Have you considered how inconsistent 21st century business mantras – like speed, agility and flexibility – are with empowerment, consensus building and large teams? There's no way you can move fast when you have to ask everyone what they think. I know, this is heresy. Twentieth century management gurus hate the idea of command and control style management in the 21st century. But hierarchical management with opportunities for participation based on merit (not golf handicaps) makes more sense in highly competitive arenas than consensus management. And it gets worse. Here are some assumptions about how to make organizations work, especially the one run by the CPO:

- **Disband Large Teams** - I realize that we've been talking about teams, bonding and collaboration for years. But how many of us really enjoy those off-sites intended to get us to relate to each other? Look, the more people working a project the more likely it is to fail. At the very least, it will cost a ton of money, much of which goes for meetings coordination.

Keep it small, keep it manageable. Big projects can be broken up into small pieces. As we've already discussed, over 75% of all uber-projects fail, so there's reason to re-think how we do things.

- **Inflate Flat Management Structures** - Whose idea was it to empower everyone? If we take this to its logical conclusion it means that everyone gets veto power over everything. Successful experience should rule the day. The last thing you want to do is empower unintelligent, inexperienced people with bad intentions. You know who they are and where they live. Keep them away from all of the important discussions. And re-think the way you manage generally. As discussed above, I still like controlled access hierarchical structures, where smart, experienced professionals make the big decisions after listening to smart colleagues who've earned a seat at the table.

- **Hire Only Partners** - Some of my best friends are consultants, but before hiring one you really have to make sure you know **precisely** what you want them to do and you understand the structure of their direct and indirect vested interests in the outcome of the work they do. One of the best acquisition practices you can adopt is shared risk: if your vendors won't share risk then you shouldn't share your money with them. It's also usually a good idea to hire honest brokers to look over the shoulders of the mainstream consultants and vendors you use, honest brokers who, regardless of the advice they provide, cannot make a dime after their gig is over.

- **Measure Everything** - I really hate this. It's almost like exercise: we do it because we have to, not because we like to. But if you don't measure things – like assets, processes, people – then you have no way of knowing how you're doing or benchmark yourself against the competition. Without empirical data, you fly blind.

- **Least of Breed** - This one will breed lots of controversy, but it makes more sense to commit to a few hardware and software vendors than to anything more than several of them. **Yes, this is an argument against best of breed in strong favor of single or double sourcing strategies.** Why? Because of the complexity of our computing and communications environments and because emerging business requirements call for scalability and agility we have to increase the chances for the successful integration and interoperability that reduces complexity and fosters agility. Look around your company to see just how much variation you're paying for and how much money you're leaving on the table. OK, so nobody's perfect, but do you really want to have three database management environments just to prove a point about independence?

- **Re-Think Relationships** - I know, here we go again. But this time there are some major differences in the maturity of both business modeling and technology effectiveness. Ten years ago business models were linear and sequential, now they're dynamic, collaborative and continuous. Twenty years ago most computing and communications technology barely worked. We're now at a very different business technology place. If we're now talking about collaborative, integrated and continuous business then we're no longer talking about business requirements thrown over the fence to eager or not-so-eager technology professionals who have to interpret what the requirements mean, but rather a holistic approach to business technology convergence that renders most of today's reporting relationships obsolete. Very simply, if your collaborative business gurus are at arms length from your technology integration gurus, and your organizational structure endorses the distance, then you have a fundamental problem in how you do business. It's time to re-think all this.

Are you still speaking to me? I hope so. These somewhat harsh views were born from some pretty disgruntled parents who eventually subscribed to tough love management.

All of this works – if you agree – for a centralized organization with one or a couple lines of business, but what if there are multiple lines? There are two ways to go, and after my list of things I'd do tomorrow, I'm curious about the one you select.

Decentralized organization #1 appears in Figure 28. It's what I call the light vise-grip structure. It keeps the self-contained structure but wraps it in enterprise-controlled strategy, operations and processes. Line of business CEOs like this structure because they have relative autonomy; they dislike it because when things go bad there are few parts of the system outside of their control that they can blame. Enterprise CEOs are OK with the light vise-grip structure, especially if they have capable line of business CEOs. They're unhappy with the structure when their CEOs begin to lose focus and they have relatively few ways to reign them back into the overall mission.

What's the next option? I call Figure 29 the tight vise-grip model because the enterprise strips the lines of business of leverage-able, scalable activities including especially all of the support and enabling technology activities. This structure gives the enterprise lots of power, but it also leverages resources

Figure 28. Decentralized organizational structure #1: The light vise-grip

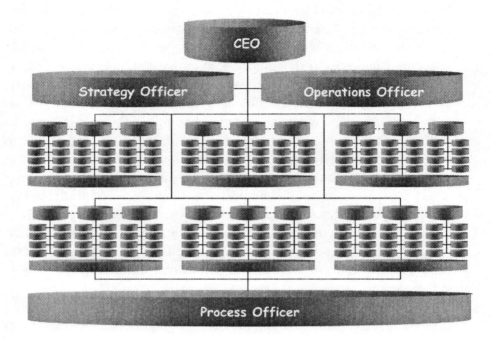

across multiple lines of business. Some CEOs of centralized lines of business might not like the structure, though some might prefer it. It depends on the culture, the vertical industry and the condition of the company.

Big surprise: I'm not crazy about the light vise-grip structure. It offers too many temptations for lines of business to wander off on their own. It also leaves a lot of enterprise leverage on the table, leverage that could reduce costs and distractions for Collaboration Officers. I like the tight vise-grip structure because it's got good economies of scale and allows the Collaboration Officers to do what they do best. In this structure, "corporate" really is there to help.

Let me tell you how **not** to achieve collaboration/integration organizational success:

- Don't appoint a collaboration "czar": czars are temporary aberrations, or at least that's how history has treated them. Czars connote a crisis of some

Figure 29: Decentralized organizational structure #2: The tight vise-grip

proportion that requires the perception of a disproportionate response that by definition burns out over time.

- Don't appoint an integration "czar."
- Don't separate – for any purposes – "business" from "e-business."
- Don't do "one-offs," exceptions to the rules – under any circumstances.
- Don't under-fund enterprise activities.
- Don't keep roles and responsibilities vague.

The Special Case of Innovation

Who's in charge of tracking business technology trends in your company? Lots of places have in-house gurus but very few have created formal positions to track the major trends that impact their companies. I've always found this amazing given the pace of technology and business change. Maybe it's time for all of us to rethink our business technology watch strategies.

So how do you identify the collaborative business models and integration technologies most likely to keep your company growing and profitable? The explosion in technology has changed the way you buy and apply technology and has forever changed expectations about how technology can and should influence your connectivity to customers, suppliers, partners and employees. What you need is a business technology investment agenda that helps you identify the technologies in which you should invest and those that get little or none of your financial attention.

The agenda ultimately must be practical: while blue sky research projects can be lots of fun (especially for those who conduct them), management must find the technologies likely to yield the most growth and profitability, not the coolest write-up in a technical trade publication.

As always, it's essential that you define the business models and processes that you'll pursue over the next two to three years. Some might argue for a longer lens, but it's tough enough to extrapolate out even two to three years. These models and trends will provide the anchor for assessing emerging business technology trends.

The purpose of business technology monitoring and assessment is to develop lists of high potential business models and technologies. Hit lists are excellent devices for rank-ordering and screening technologies. They also focus attention on specific business technology opportunities.

The most promising business models and technologies should be piloted. The purpose of the pilot is to determine if the technology will cost-effectively solve problems that to date have proven difficult and expensive to solve. Pilot projects should be real projects. They should have project managers, schedules, milestones, and budgets. They also need dedicated professionals to objectively determine where the promise really lies.

Pilot projects should not last long: a pilot project that requires six or more months to yield the classic go/no go result is much less likely to succeed than one that yields an answer in 60 days. Succeed – or fail fast – and cheaply. In fact, if you institutionalize the piloting process, your ability to attract funds to conduct technology pilots will correlate to how quickly you've delivered results in the past.

All of this needs to keep happening: an effective business technology watch strategy continues forever. You need to model your business models and processes continuously as well as the technologies likely to enable them. Some companies institutionalize the process in the form of in-house R&D labs, "skunk works," or "incubators." You should do whatever's likely to work.

Innovation has a big appetite. It requires lots of fuel, patience and commitment. The big technology vendors understand this. Billions are spent every year to gain competitive advantage. We can learn a lot from their innovation strategies.

I was recently invited to the Pentagon to talk about the adoption of new technology. It was part of the government's effort to (once again) "transform" the way it does business. My role was to help them think about how to introduce new technology to old problems, processes and decision-makers. I agreed to go because I wanted to draw some distinctions among technology concepts, working prototypes and technology clusters.

So what is grid computing? Is it a technology concept, a prototype technology or a whole technology cluster? What about voice recognition technology, semantic understanding and the Segway? Are they concepts, emerging technologies are part of larger technology clusters?

I got to thinking about all this because I recently did a "content analysis" of a number of technology trade publications and turned up no less than 30 "technologies to watch." Content analysis is a technique that identifies trends by counting the frequency of mention: if something's mentioned a lot – like Web Services – then it ranks high in the analysis. Lots of technologies get mentioned a lot – which is why there are 30 or so "to watch" – but let's be honest: there's no way all of them deserve our attention – or our money. But what are the technologies that matter?

I segmented technologies into concepts – ideas like "real-time computing," emerging technologies, like wireless networks, and technology clusters that include real technologies plus infrastructure, applications, data, standards, a developer community and management support. I then made the argument that technology impact was related to concepts, technologies and clusters, that concepts are wannabes, prototype technologies have potential and mature technology clusters are likely to have huge sustained impact on the way to do business.

I then mapped a bunch of the technologies-to-watch onto an impact chart and discovered that many of the technologies about which we're so optimistic haven't yet crossed the working prototype/technology cluster chasm – indicated by the thick blue line that separates the two in Figure 30. Technologies in the red zone are without impact. Those in the yellow zone have potential,

Figure 30. Technologies, impact and the chasm

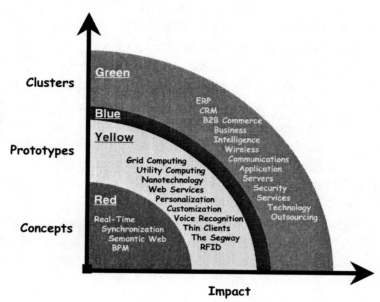

while those in the green zone are bona fide. The chasm is what separates the yellow and green zones.

The essence of all this is that technologies will have limited impact until full clusters develop around them consisting of all of the things necessary for technologies to grow, all of the applications, data, support, standards and developers that keep technologies alive and well over long periods of time. Figure 30 also suggests that it's too early to tell if many of the technologies-to-watch will become high impact technologies, that is, if they will cross the chasm. Real-time synchronization, business process modeling, grid computing and utility computing, among others, may or may not yield successful prototypes – which may or may not evolve into full-blown clusters.

So how did I do with the heavy hitters at the Pentagon? A few of them thought I was Geoffrey Moore, the co-author of the now classic **Crossing the Chasm**. Some others thought I worked for the technology vendors who had actually crossed the chasm, and a lot of them thought that the whole notion of clusters was too restrictive, that technologies – even if they were bogus – needed nurturing. When I said that I thought such an approach could prove to be very expensive, they reminded me that their job was to invest in high risk/high payoff technologies, not to invest in technologies that were definitely going to work.

On the train back from Washington, I finally figured out where all the money goes – and how companies can save some of the money the government spends:

- Mostly buy clusters
- Occasionally invest in prototypes
- Enjoy (but don't buy any) concepts

Or put another way, unless you're in the technology business, don't be an early adopter, a pioneer or live on the bleeding edge. It's too expensive (unless you work for the government).

Is this a good way to segment technologies? I think it helps categorize the phases technologies go through and helps us avoid investing too early in technologies that haven't proven themselves. Stay in the green zone and, if you have to wander, don't leave the yellow zone. The red zone's a money pit: track the returns here on other people's money.

Business Technology Organizational Convergence

Well, here we are. Over a hundred pages into the convergence pitch and we're finally talking about organizations (and we still haven't talked about people). The key is the nagging relationships among collaborative business speed and agility and organization structure and management styles. The whole notion of organizations dedicated to what they do – and nothing else – is obsolete in a converging world. The key functions – collaboration, support and enablement – are all on the same level, but the overall strategic and operational responsibility for these units is coordinated by a central command structure. Even in companies with lots of lines of business, the command structure should remain pretty much the same.

Innovation deserves very special attention if for no other reason than the velocity of business and technology change and therefore the velocity of convergence.

Who Wants to Go First?

The CEO ...
"I do ... I want to go first ..."

The COO ...
"Sure you do ... you love this command and control stuff ..."

The CFO ...
"Not as much as me, I'll tell you that right now ..."

The CKO ...
"This is pretty rough stuff ... are we sure we like this approach?"

The CIO ...
"We like it ... we really do ..."

The CEO ...
"Good, because you and the CTO are history ..."

The CIO ...
"I thought we were just getting new titles?"

The CFO ...
"I don't think you read the memo ..."

Chapter VI

The Management Conversation - It Still Needs To Make Sense

Here's what we'll discuss here:

- Measurement – or do you know where your computers, processes and skeletons are? Without benchmark data it's impossible to converge anything: measure or fly blind.
- The standardization of your computing and communications technology. If people want to buy non-standard, non-supported hardware or software make them pay for their own support. Watch three infrastructure levels: access, coordination and resource, and measure everything so you know what works and what doesn't, and what things cost.
- Outsourcing, or the love/hate relationship you should have with the people inside that are good/bad and the people you hire from the outside who are good/bad/expensive/cost-effective, and why you should outsource only to partners willing to share risk.
- Funding, or figuring out who pays for what at your company, and dealing with the inevitable conflicts between the "enterprise" and business units.
- Return-on-investment (ROI) and total-cost-of-ownership (TCO), the "I-see-no-compelling-reason-at-all-to-fund-this-project" twins – who really are your friends (so long as they stay on their medication).

- Business Cases, or the "OK-I-guess-we-have-to-do-this-one" police.
- Project/Program/Portfolio Management – the once-and-always business technology management champ.
- Governance – or the way we decide how to organize ourselves, regulate ourselves, comply with regulations and legislation and otherwise tell ourselves what to do.

You Still Think So?

We've talked about organization structure in Chapter V, so let's talk here about some business technology convergence management priorities. Let's stipulate that the practice of business technology management is seldom as "professional" as it ought to be. On the one hand, we know exactly what to do – and not do – but on the other hand, we seldom do what we know is right. For example, we don't always develop solid business cases for large technology investments. Nor do we measure the variation in our computing and communications technology environments in order to standardize as much gear as we can. Nor do we insist upon project management best practices. Our total-cost-of-ownership (TCO) data is often incomplete and we tilt with windmills on the importance of return-on-investment (ROI) calculations, not to mention all the flavors of ROI and TCO methods we like. We don't require our vendors to share risk and our partner management skills are far from adequate. Why am I whining? Remember that the probability of a successful technology project is somewhere around 25%.

You still think you can manage?

How Much Do You Know?

Let's start with the basics. How much do you know about your business technology? Why is it so important to know what you have, how you get it, what you do and how you keep it all going? Years ago someone asked me where we disposed of the machines that we upgraded out of existence. I did not know the answer, so I began asking around. Turns out that we actually "stored" some of the obsolete machines in the drop ceilings of our offices. Intrigued, I then asked about the existing inventory of stuff: "somewhere between 2,000 and 3,000 servers, somewhere between 5,000 – 7,000 laptops and around 40,000 – 45,000 desktops – but we don't know where they all are." And this didn't include the machines in the ceiling.

Then the conversation turned to processes like software development, business scenario planning and sourcing, and the answers were all over the place again. Obviously there was little or no measurement in the organization, and the culture – for whatever reasons – did not support the continuous measurement of hard and soft business technology assets. In fairness, human resources **did** know how many people worked for the company – but that, I assumed, was because they always had lots of time on their hands.

The area of measurement is fascinating because everyone thinks they measure lots of things and almost no one does. We think we know what we have, who works for us, their skill sets, the applications they use, how happy our customers are and the rate at which we're growing. But we don't. In fact, most organizations have barely inventoried their assets, their business processes or the business and technology outcomes that should matter the most. In bull markets when customers and capital are plentiful, hardly anyone gives a damn about measurement, but in bear markets everyone wants to know everything (especially where to cut expenses). Obviously, measurement is essential to management regardless of the capital markets.

Here are 25 questions. See how many you can answer:

1. How many PCs/PDAs/servers does your organization own?
2. How many PCs/PDAs/servers has your organization purchased over the past 24 months?
3. Where are the PCs/PDAs/servers?
4. How many applications do you support?
5. How well do they perform?
6. How many platforms are they running on?
7. How many architectures do they represent?
8. How many networks are in your organization?
9. Who owns the voice and data contracts you're recently signed?
10. How many IT professionals – including full- and part-time consultants – work for you?

11. What are their skill sets?
12. What are the knowledge and technology gaps that threaten your productivity most?
13. Who are your strategic partners?
14. How and why are they contributing to your tactical and strategic goals?
15. What are your core processes?
16. Do you routinely do risk management?
17. Do you have a standard systems analysis and design life cycle?
18. Do you have standard acquisition contracts?
19. How many procurements have you bid in the last 24 months?
20. What is your systems services quality of service?
21. Do you have service level agreements in place?
22. Who owns them?
23. Who's accountable for what's in your service organization?
24. Do you have ROI data?
25. Do you know the total cost of ownership (TCO) of your desktops, laptops, PDAs, applications and networks?

How Many Could You Answer?

If you can't answer these kinds of questions, you have a measurement problem. (A good score is 15. You're toast if your score was 10 or below, and golden if you were 20 or above!)

As always, the answers must have business purpose. The reason why it's useful to know how many PCs were purchased last year is to track trends – and the trends provide insight into relationships among professionals, productivity and costs. Collecting lots of data for no particular reason is silly. Measurement data is useful only if it's leveraged onto convergence decision-making.

This conversation will do two things. First, it will identify – in true "book of lists" fashion – what the measurement issues are, and secondly, it will offer some "if/then" rules for inferring the significance of the measurement data. Hopefully, the information provided here will jumpstart your measurement efforts and address some of the political challenges you'll inevitably face.

The previous conversation focused on specific business technology areas (such as applications, communications and organizational structure). Measurement is essential to all of them, since it's impossible to modernize your applications portfolio without knowing what's in the existing one, or to determine if you have enough communications bandwidth if you don't know how much is enough, or define and enforce standards if you're unsure about the variation in your environment. Measurement is key, but the effort to measure should not exceed its benefit. This "heads-up" is to those who may have heard that measurement can only be accomplished from the cockpit of a Gulfstream V or the front seat of an S-Class Mercedes. Piper Cubs and Chevies can also get the job done.

Here are some pieces of the measurement challenge.

First, the ultimate metrics are business, not technology metrics, though the latter should be derived from the former. This means that technology metrics – like the cost per MIPS (millions of instructions per second) is only meaningful with reference to throughput and cost-benefit requirements. Similarly, knowing how often you upgrade desktop or laptop PCs only matters if you're significantly out of line with your industry's best practices and/or you cannot determine why you're upgrading machines at all. While there may some very good reasons why you're upgrading at a much faster rate than your competitors, you really need to know what they are before you continue spending.

Figure 31 can help with a current baseline assessment and with the prioritization of measurement requirements.

Figure 31. A measurement requirements and planning matrix

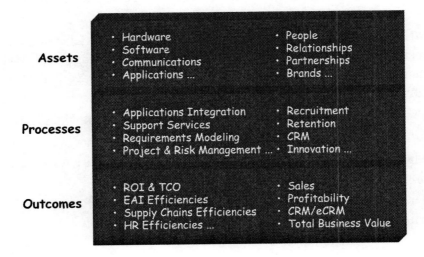

Let's start with assets and let's define them very broadly. As with the other two measurement areas – processes and outcomes – we'll focus on relatively "hard" and "soft" metrics. Here's the list of hard assets you should measure (most – though not all – will apply to you).

If you're in a large organization – especially a publicly owned one – your auditors may be able to help you gather this information. If they don't have it and haven't asked you for it – yet – then brace yourself, because the judgment day is approaching when they'll expect you to have all of the numbers.

"Hard" Assets
- #/Location/Assignment/Age of Desktop Personal Computers
- #/Location/Assignment/Age of Laptop/Notebook Computers
- #/Location/Assignment/Age of Personal Digital Assistant (PDA) Devices
- #/Location/Assignment/Age of Servers
- #/Assignment of Local Area Networks Under Your Control
- #/Assignment of Wide Area Networks Under Your Control
- #/Assignment of Virtual Private Networks Under Your Control
- Description of Network Topologies in Your Organization & Under Your Control
- #/Location/Assignment/Age of Mid-Range Computers
- #/Location/Assignment/Age of Mainframe Computers
- #/Location/Assignment/Age of Storage Devices
- #/Location/Assignment of Desktop/Laptop/PDA Applications & Licenses
- #/Location/Assignment of Utility Applications (Such as Change Management, Configuration Management, Requirements Management, Maintenance, Testing and Related Applications Across All of Your Computing Environments)
- The Original Equipment Manufacturer (OEM) Brands of Your Hardware Suppliers; the Percentages Across Vendors
- The Distribution of Software Vendors in Your Software Asset/License Pool
- The #/Location/Ownership/Age of Applications by Platform: Mainframe, Mini-Computer, Client-Server, Desktop

- The #/Location/Ownership/Age of Applications by Architecture: Single-Tier, Two-Tier, Three-Tier, Multi-Tier, N-Tier

"Soft" Assets
- # of People in Your Organization
- Their Educational Backgrounds & Professional Experiences
- Their Current Applicable Skill Sets
- The Mapping of Those Skill Sets Onto Your Requirements; Skill Set Gap Analyses
- The Salaries & Bonuses (and Other Parts of Compensation Packages) of Your Professionals and Support Staffs
- The Intra-Corporate Relationships and Partnerships
- Your External Alliances and Partnerships
- Your Brand(s)
- Your Goodwill
- Your Professional Reputation

Measurement is pointless without purpose. Here are some of the high-level rules of thumb that extend from asset measurement data. While the list below is by no means exhaustive, it's representative of the kinds of decisions you can make from measurement data:

- Data About the Age and Configuration of Your Desktops Should Inform Your Desktop Upgrading Strategy. Over Time, You Will Learn about Optimal Holding Periods for PCs (and Other Devices) that Will Permit You to Only Upgrade When Necessary (and Thereby Avoid the Costly Churning from PC to PC for No Empirical Reason).

    Rule: **If** Your Need for Newer Versions of Primary Desktop Applications and the Platforms They Run On is Minimal or Manageable, **Then** Institute a 36-48 Month "Holding Period" for Desktops, Laptops & PDAs.

- Data About the Number, Age, Architecture and Platform Base of Your Applications Will Provide Insight into the Support Necessary to Keep Them Going. If You Discover a Lot of Different Architectures, Lots of Different Platforms and as Many Old Applications as Newer Ones, Then You're Probably Paying Too Much for Support. The Data Would Suggest the Cost-Effectiveness of Reducing the Platform/Architecture Variation in Your Applications Portfolio.

  Rule: **If** Your Applications Platforms and Architectures are Highly Varied, **And** Your Goal is to Reduce Support Costs, **Then** Reduce the Variation in Your Computing and Communications Environment.

- The Rates of Adoption of New Computing Devices Will Tell You a Lot About Where Your Business Processes are Going. For Example, If You Find that You're Adopting Laptops and PDAs at a Much Faster Rate than Desktop PCs, then You're Probably Moving Toward a More Distributed Workforce and Customer/Supplier Network that Will Stress Your Remote Access, Security and Data Storage Capabilities.

  Rule: **If** the Rate of Laptop Adoption Outstrips Desktop PC Deployment by 1.5:1 **Then** Check Your Remote Access and Security Services Capabilities and Prepare for Additional Investments in Distributed Remote Access, Security and On-Going Support.

These "rules" suggest how to convert dry, dead data into serious cost-savings and efficiencies. The conversion is key to the success of your measurement strategy. Insight into your hard and soft assets can lead to all sorts of discoveries, such as huge gaps between what you need to do (like integrate applications) and your existing skill sets (you have no applications integrators). You might find that your brand and external image are inconsistent with your new business models and process, or that your partners (suppliers and distributors) are confused about where your business model ends and theirs begins. All sorts of great and horrible news about your assets is just waiting to be discovered.

Processes represent a special kind of measurement challenge, since they're so often extremely soft and often ambiguous. The processes you should be concerned about include – at a minimum – the following:

"Hard" Processes
- Your Systems Analysis and Design Processes (Your Life Cycle Methodology)
- Requirements Management Processes
- Risk Management Processes
- Project Management Processes
- Process Adoption/Sustainment Rates
- Service Level Agreements (SLAs) and Success Rates
- Hardware and Software Acquisition Processes
- Hardware Disposition Process
- Asset Management Process
- Network and Systems Management Processes
- Vendor Management Processes
- Vendor Selection and Management Process
- Help Desk Processes
- Security Authentication Processes
- Security Authorization Processes
- Security Administration Processes
- Disaster Recovery and Business Resumption Processes
- Data Base Administration Processes
- Knowledge Management Processes
- Standards Setting Processes
- Standards Governance Process
- Business Technology Audit Processes

"Soft" Processes
- Your Human Recruitment Process
- How Effective Has It Been (Measured by # of Recruits & the % that Have Stayed Over Time)?
- Your Employee Performance Review Process
- Your Jobs/Opportunities Description/Classification/Posting Processes
- Your Benefits Administration Processes

- What are the Processes that "Touch" Your Customers?
- How Effective Have They Been (Measured by Customer Service Data)?
- What Processes Have You Implemented to Stay Technologically Current? Is There an Internal R&D/Innovation Process? How Many Internal Proposals Have You Received? How Many Have Been Funded? How Many Have Been Successful?

Where's your industry? Let's speculate that your industry may be crazy, so benchmarking its performance may not make perfect sense. But given it's where you work, it's still good data to have. Given how vertical industries are converging, it also makes sense to look at vertical industries close to your own for insight into assets, processes and outcomes trends and best practices. Investments here are worth the money.

So what's the deal here? Well, you need to know where you are, where your competitors are and where your industry's going in order to make informed decisions about business technology. I don't think you need a full orchestra here. A quartet that you can count on to deliver really good music will suffice. Why no fanfare? Because I'm still recovering from TQM.

Variation's Your Enemy

The area of standardization is bathed in emotion. Nearly everyone in your organization has an opinion about what the company should do about operating systems, applications, hardware, software acquisition, services and even system development life cycles. Everyone. Even the people who have nothing to do with maintaining your computing and communications environment will have strong opinions about when everyone should move to the next version of Microsoft Windows and Office. In fact, discussions about standards often take on epic proportions with otherwise sane professionals threatening to fall on their swords if the organization doesn't move to the newest version of their favorite application.

It's likely you've heard references to return on investment (ROI) and the total cost of ownership (TCO) every time the subject of standards comes up (we'll talk more about TCO and ROI later). Just so there's no misunderstanding here, there is no question that environments with less-rather-than-more variation will save money and increase efficiency. Or put another way, you have a choice here: you can be sane or insane.

What does business technology management require here? Standards are a second-order business driver. Most businesses don't associate standards-setting with business models, processes, profits or losses. Whether the environment has one, five or 20 word-processing systems has seldom been associated with business performance. It's hard to link homogeneity with sales. But the fact remains that expenses are clearly related to sales, and standards are closely related to expenses.

What else does business technology management require? Flexibility – and here lies the only real argument against standards. If your environment doesn't support the business computing or communications processes the business feels it needs to compete, there will be loud complaints. Business managers want to compute and communicate competitively. Standards are often perceived as obstacles, not enablers.

Try out these standards questions:

- How varied is your current environment? How do you know?
- Do you know what it's costing you to support a highly varied environment?
- What is your organization's tolerance for governance of any kind? For standards governance?
- Who's in charge of standards in your organization? Who's not?
- Is there a business buy-in to the concept of standards and to the cost-effectiveness of standards?
- Has your organization been audited for its compliance to standards? The result?
- Do you have standard desktops, laptops and PDAs?
- Do you have a standard communications architecture?
- Do you have a standard applications architecture?
- Do you have standard hardware and software acquisition practices? Do you procure centrally?
- Do you have standard design, development and project management standards?

The answers you give to these kinds of questions will reveal a lot about your philosophy about standards, standards responsibility, authority and accountability – and whether or not your chances of standardization are high, low or miserable.

If we've learned anything over the past few decades, it's that standards are as much about organizational structures, processes and cultures as they are about technology. The ability to actually control computing and communications environments through thoughtful governance policies and procedures will determine to a great extent how standardized organizations become. We've also learned that the more you succeed, the less you pay. And, yes, there's the wimp factor. Lots of companies say that have standards and tough governance policies but they really don't. Walk around your company. Do you see three or four different desktop PC manufacturers? Do you support IBM, Oracle, Sybase **and** Microsoft databases?

As always, everything needs to sync with your collaborative business strategy. Key here is governance and the processes that make standards management effective. Without serious support for a standardized environment, you're toast. Clearly, less variation in your platforms, applications, architectures, acquisition and disposition practices, and life cycles will reduce your costs. And as always, you need to focus on what the environment should look like in the next two to three years.

The following figure will help you implement a standards strategy. It offers cells in a matrix that you can use to identify, define and prioritize requirements (or spend some time in if you choose to be insane).

Note the distinction between the enterprise and the business division or units. This is a killer distinction since it determines ultimately whether you succeed or fail. If you're in a strong centralized organization then your chances for success are much, much higher than if you're in a decentralized organization with a weak enterprise group responsible for infrastructure. Put another way, it you're in an organization that has a central CIO whose job is really a "Chief Infrastructure Officer" and whose charter is full of authority holes then you're not likely to reduce variation in your environment. In fact, you're likely to find yourself suiting up for one standards crusade after another.

The organizational structures that have the greatest chances of success are either a strong centralized organization or a decentralized one that has unambiguous separation of duties, with the infrastructure usually belonging to the central group. This later model only works when there is a buy-in to standardization and when buy-in begins to weaken central management steps to re-establish standards authority.

The organizational structures we discussed in Chapter V can help here. The light and tight vise-grip structures both place standardization power in the hands of the Process Officer, who has both the **responsibility and authority** to enforce corporate standards.

Figure 32 requires you to look at your governance (more on this a little later) and processes, your platforms, your primary software applications, your architectures, your acquisition and disposition standards, as well as your life cycles. The objective of these assessments is to reduce variation as a means to save money and preserve flexibility. The figure also requires you to realistically determine your organizational structure's relationship to standards-setting. If you're decentralized, then you have some serious governance work to do. If you're centralized, then you'll have fewer religious wars over standards. The figure asks you to think about the enterprise versus the divisions or business units and proceed accordingly.

OK, now let me rant for a while on this subject. Let's start with our attitude toward choices. We love lots of choices when it comes to restaurants, cars and travel destinations. We love all the different ways we can have coffee. We love choices. We're conditioned from birth to love choices. We also like change. If we do something over and over again, we get bored. Change is our friend: I mean, let's not go there for lunch **again**!

Figure 32. A standards requirements and planning matrix

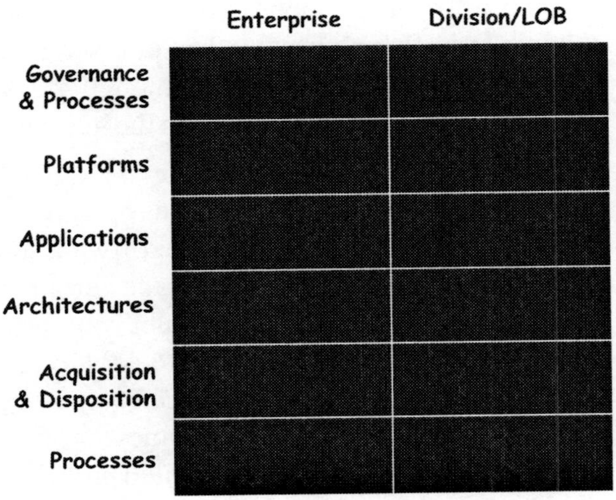

	Enterprise	Division/LOB
Governance & Processes		
Platforms		
Applications		
Architectures		
Acquisition & Disposition		
Processes		

There are lots of technology vendors out there. Probably 20 companies you can buy PCs from – not to mention all the white box manufacturers or distributors of the same stuff. Some of these machines are really cool – thin, aesthetically pleasing, even artful. Some are real fast, and some have real big, flat, bright screens. Some are "rugged," and some are really lightweight. See where I'm going with this? The same is true of software, networks and databases. They're all as different as they are the same.

Everyone also has opinions. If you bought a Mercedes, then you like Mercedes-Benz automobiles and you want everyone to know it. If you bought a Sony Vaio PC you want everyone to admire the wisdom of your excellent decision. A CEO is hired away from a competitor that was committed to Oracle databases and enterprise applications. In fact, the decision to migrate to Oracle was driven in the old company by the new boss. But the new company is an IBM database shop running SAP's enterprise applications. When the new CEO arrives everyone hears about how the old company dumped their IBM/SAP applications and systematically moved to Oracle – and of course how wonderful it all turned out. Here we go again.

Sorry to sound dictatorial about variation, but not only should you reduce – or if possible eliminate – variation, but you should also stay as mainstream as possible. This is an argument for big, powerful, sometimes arrogant single-source standardization, and against best-of-breed, mix-and-match technology acquisition strategies. Sorry, little vendors. But remember that the field's prototyping phase is over. It's now time to get serious about your business technology and to reduce distractions that come from high hardware, software and communications variation. Non-standardization is just plain stupid. The more non-standard you are, the more money you're leaving on the table (and taking from your bottom line). Remember also that some variation is inevitable. It's impossible to find just one or two vendors that will satisfy all of your technology requirements, but try like hell to keep the number as low as you possibly can.

One of the questions you'll have to answer concerns open versus proprietary software, and whether or not to make open source software standard in your company. There's a grass roots movement to make certain infrastructure software – like operating systems – "free." No doubt you've heard about the Linux operating system. It's cheaper than Windows or Unix, comes with lots of free development and management tools, can often outperform proprietary software with less hardware, and is actually somewhat more stable that proprietary software because of the speed with which problems are detected and fixed by the online, open source army of developers. Some open source software is supported by large vendors, like IBM, HP and Oracle, as well as smaller vendors with interesting names like Red Hat and SuSE (now part of Novell).

But open source software is far from perfect – and ultimately, at least in my mind, nets negative. Here's why. There's a ton of popular software that won't run on much of the open source stuff, not the least of which is Microsoft Office. Lots of the open source stuff is very geeky. You need a bunch of extremely technical people to make it work well, so technical in fact that you can be sure that they won't corrupt applications or databases through unrestricted access to source code. There are also intellectual property issues when proprietary software is mixed with open software and distributed to other users: complicated laws and regulations dictate whether or not the software stays open and whether your proprietary software stays proprietary.

Look, you can tilt at windmills, try to run your company without any Microsoft software, and rely completely on open and free stuff, or you can get real and learn how to optimize the hardware, software and interoperability of as few vendors as you can.

This also applies to technology standards like XML, 802.11a/b/g and CFAR, standards that support integration, communication and supply chain management, among other emerging dominant standards. Here again you should stay mainstream. Watch the major vendors to determine which way the standards winds are blowing. Don't be bleeding or leading edge here: there's no reason to gamble, no reason to make expensive bets on wannabe technology standards. You have no reason in the world to speculate. Of course if you're in the standards business then – as Captain James T. Kirk used to say – "risk is (y)our business." (Yes, I know that the "T" stands for Tiberious.)

Web Services technology standards, for example, are emerging but more slowly than some would like. As suggested in Chapter IV, track them carefully. There's even a group – The Web Services Interoperability Organization – dedicated to advancing common standards (www.ws-i.org).

The bottom line?

- Reduce the variation in your technology environment as much as you possibly can.
- Watch the big players for the "right" technology standards.
- Stay with mainstream hardware, software and communications providers.

- Viciously quell any standards revolutions that break out and publicly hang the leaders.
- Sit back, save money and reap the rewards.

Can We Really Do This?

Are you outsourcing yet? If you're not, then you're in the minority. Most everyone outsources some part of their business technology operation for all sorts of good – and occasionally very bad – reasons. There's a reason why the technology services industry is clipping along at over $1B per day in the U.S. alone. More and more companies have discovered the benefits of outsourcing compared to the recruitment and maintenance of large internal business technology staffs. In the early years, we all thought outsourcing was about saving money, but then we discovered the truth: outsourcing is not only about saving money, it's about re-routing money from non-core to core activities.

Acquisition strategy is – when all's said and done – about whether or not you should build and maintain a large internal technology staff. All of the books, articles and seminars about core competencies are – in the final analysis – about shedding processes. The core competencies drill is critically important to acquisition effectiveness. As your collaborative business evolves, you need to ask tough questions about maintaining the in-house activities you've supported for all these years. Remember that the assessment is not just about cost.

You get the idea. The key questions have to do with defining your core collaborative business purpose and then matching all of the activities to in-house versus outsourced alternatives. Once you've determined what makes sense, it's possible to step back and assess the kinds of technologies, processes and services that might be outsourced. But just in case you think that all roads lead to outsourcing, make sure that you objectively assess the impact outsourcing will have on specific collaborative business models and processes. It may be that certain activities should stay well in-house. Figure 33 presents the range of services you either do in-house or might outsource. Which ones do you do in-house and which ones do you outsource? Why?

The discussion here is about structure and form, not about whether outsourcing will play some role in your business technology acquisition strategy.

Figure 33. Outsourcing candidates

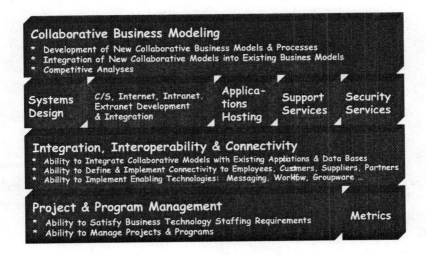

You have a number of outsourcing options:

- Combine outside vendors with your own. Sometimes called in-sourcing or co-sourcing, this model can be very effective if structured and managed properly.
- Completely outsource segments of your technology mission, such as data center or call center management, but keep others in-house. This option can also be effective, especially when there are clearly defined areas that you do well and those they you do poorly – and when there's no ambiguity about what's core and what's not.
- Completely outsource everything to vendors who come on-site and manage your business technology resources (including machines, networks, and people).
- Completely outsource everything to vendors who "rent" hardware and software back to you.

Of course there are variations on all of these, but these four identify the primary outsourcing models you might consider.

All of these variations require that you:

- Systematically identify requirements
- Compare current (so-called baseline) costs with what outsourcers bid
- Negotiate with the vendors on price and services
- Develop clear and unambiguous service level agreements
- Make sure that management is in place to monitor the results of the work

It's strongly suggested that you seek outside help to develop your outsourcing strategy. I realize that this may sound absurd: the recommendation is that you outsource the work necessary to outsource the work, but the fact is that outsourcing is complicated and there are now consulting organizations that specialize in this kind of work. These consultants have experience writing requests for proposals (RFPs), screening the proposals and the bids, negotiating contracts, and then managing at least the initial implementation phases – but they should have no long-term vested interests in the recommendations they make. (Yeah, I realize that this should be obvious, but I just had to say it again.)

There are also some rules of thumb you might want to consider:

- Above all else, your outsourcing process should be driven by the results of your core competency assessment and you skills gap analysis. If you find that you really don't need to be in the data migration business **and** that you have no data migration talent in your shop, but that data migration is an important (though non-core) component of what you need to do, then obviously you need to outsource data migration (probably as part of some large applications modernization process).
- Make sure you know what you're doing. While evolutionary experimentation is often a good way to learn about some new process (like outsourcing) it may not be prudent. Breaking off pieces of your internal IT shop to give to outsourcers to try them out may make abstract sense but in practice may be doomed to failure. Why? Because you're likely to outsource the pieces that are the most politically correct while avoiding the really hard decisions about what's core and what's not.
- Be careful with outsourcing deals intended to transfer knowledge from the outsourcer to in-house professionals. We learned in the late 1980s and early 1990s that knowledge transfer-based outsourcing deals were

difficult to make work. Why? Because the outsourcer had no incentive to transfer knowledge and the in-house professionals resented the "training" forced down their throats.

- If you want to try outsourcing on for size, then partition a big piece of your infrastructure – like your data centers – and outsource them – completely. Develop some clear service level agreements and then monitor the hell out of the performance to see if: (a) the outsourcer can do it more cheaply and (b) better. The implied suggestion here is to outsource what you already know how to do and fully understand, not what you don't understand. And remember that just because you understand how to, for example, run a data center, it doesn't mean that it's core to your business.

- Really think long and hard about using professionals to architect your outsourcing deals. If you're a medium-sized organization or one that has had some extraordinary technology infrastructure or applications problems over the years, you might want to take a look at using an applications service provider (ASP) who will "rent" applications to your users (who can access the applications over the Internet or through a [much more expensive] virtual private network). This kind of outsourcing is relatively new but already the major systems integrators have begun to partner with enterprise software vendors like SAP to provide access to major applications. It's something to consider, especially given how tough it is to implement enterprise applications.

- The age of non-shared contracting is over. Every outsourcing deal you sign should have some shared risk built into it. If the outsourcer is unwilling to put any skin in the game then there may be a problem with the whole deal. A confident outsourcer should welcome the opportunity to jointly develop some performance metrics and then hit the metrics to get paid. These deals can take all kinds of forms. For example, expenses can get paid but a percent of profit may go into an escrow account to be paid as milestones and metrics are achieved. Regardless of the form, the principle is to share the best and worst aspects of outsourcing by aligning all of the incentives.

- Strongly consider owning requirements, specifications and designs, but not implementation or support. This rule of thumb is not inviolate but will serve you well. In a sense, owning requirements, specification & designs keeps you in control of the business technology convergence process while freeing you from (probably) non-core implementation and support tasks.

- Make sure that metrics are in place long before you sign any outsourcing deals.
- Do not sign any long-term outsourcing deals unless the deals have huge shared risk features.

It's critical that all of the analyses, assessments, requirements, baseline costs, new costs, shared risk assumptions – and especially performance metrics – get quantified.

Depending on what you've chosen to outsource, you should develop a set of metrics that will permit you to first compare what you've got now to what was the case before outsourcing and if the outsourcer's performance is up to snuff. There should also be metrics to determine if in-house professionals are performing adequately, should you decide not to outsource.

The metrics should be rolled into formal service level agreements (SLAs) that should form the basis for the outsourcing contracts you sign. Again, SLAs should be used to track in-house or outsourced performance.

The SLAs should also anchor your shared risk arrangements. So long as performance metrics have been quantified, the shared risk deals can be assessed. If you haven't quantified expectations then you'll have problems with your provider.

All of this needs to be monitored closely because dramatic change is inevitable. As suggested above, one that should be especially tracked is the new movement toward applications hosting by third-party vendors. Other trends such as shared risk, premium pricing, and related incentive structures also bear close scrutiny.

Business process outsourcing (BPO) is consistent with business technology convergence. We've been talking here mostly about technology outsourcing but the same core competency arguments apply to business processes and business process outsourcing. BPO is gaining steam. Companies are handing over payroll, human resources, training, accounts payable and administration to outsourcers willing to perform such activities often under unique contracts that involve stock and other compensation.

There are some special areas of opportunity and concern that we should talk about:

- Off-Shore Outsourcing and Recruitment
- Security Solutions Outsourcing
- Technology Utilities
- Timing

Off-shore outsourcing is getting popular again. Some years ago we experimented with sending coding assignments to India with limited success. Today, the same opportunities and risks surround offshore outsourcing. So what do you do? The appeal of course is cost savings and, increasingly, discipline in the form of elegant software documentation and improved reliability – in the case of software development – and efficient processes and excellent customer interaction - in the case of call and customer support centers. Companies have outsourced technology and business processes to a variety of companies outside of the United States in order to save money and improve quality. Does it work?

Sometimes it works great. When the tasks are well-defined and highly doable – and when you know a lot about the tasks – it can work well. But when you outsource a problem you cannot solve it's likely to stay unsolved. Here's the ugly truth about outsourcing: never, ever outsource something you don't thoroughly understand. If you build bad software systems, you'll probably execute a bad software development outsourcing deal. If your back office processes really suck, you're unlikely to find someone who can do it faster, better, cheaper – unless you hire an outsourcing expert to write the request for proposal and service level agreements. When the outsourcing deal is with a company thousands of miles away it better be well-conceived and well-oiled. Bottom line? Outsource in outer Mongolia only when you understand the processes and objectives, have a mutually beneficial but explicit service level agreement, and metrics that enable you to track performance at least quarterly – if not monthly.

Security solutions outsourcing has some unique characteristics. Large enterprises have been hammered by auditors who love to prepare letters chastising their lack of internal and external security. Smaller organizations are scared to death of viruses and worms, while the surviving dot.coms are painfully aware of their need to make sure that Internet-based transaction processing is secure. As online business-to-business (B2B) and business-to-consumer (B2C) transactions increase, organizations will have no choice but to deploy digital signature, public key infrastructure (PKI) and perhaps even biometric authentication technologies. This stuff is getting very complicated very fast.

So how should you migrate from where you are today to where you'll inevitably need to go? Why should you migrate at all?

Look at all of the pieces to the security puzzle and how deployment is dependent upon the integration and interoperability of a ton of technologies, products and services. Some of these technologies, products and services your in-house security people can handle, but some are well beyond their capabilities. Who owns security technology/products/services integration? Who makes sure that all the technologies, products and services interoperate? Key here is the development of a comprehensive security policy that defines a security architecture that describes how authentication, authorization, administration and recovery will occur inside and outside of the corporate firewall. But who should own the pieces of what should become your security **solution**?

Regardless of how many disparate security pieces you have today, you probably have too many. If you have more than one redundant or overlapping security service level agreement you have too many. And if you've distributed security accountability across your organization, you're security efforts are diffuse at best – and dangerous at worst.

Since just about every business on the planet will have to integrate their traditional business models with models that exploit Internet-driven collaboration, bulletproof security will become a transaction prerequisite. Since just about every business has under-spent on security, additional resources will have to be found to solve the inevitable problems that distributed business models will create. All of the authentication, authorization, administration and recovery problems will have to be solved by stitching together a variety of technologies wrapped in products and services. What will these technologies be? What products will you use? How will you support them? Can internal staffs cope with the changes?

Some advice. Unless you're a security products or consulting company, get out of the security (and privacy) business. It's time to consider outsourcing security to vendors who can provide reliable, integrated, interoperable solutions. But this advice does not extend to the specification of security requirements or the development of security policies. It's always prudent to own requirements and specifications – strategy – and optimize the implementation and support of that strategy – tactics. In other words, it makes sense to in-source strategy and outsource tactics.

Remember that success here is defined only around **solutions** – the integration of technologies, products and services that work together as seamlessly and efficiently as possible. You need a single point of accountability who really gets your collaborative business. You need killer security requirements analysts who can specify security policies and architectures, and you need professionals who can manage the implementation of those requirements through the creative synthesis of security technologies, products and services.

Your solution will be a hybrid that integrates some existing technologies and processes with a new set of technologies, products and services. The technology architecture must be flexible and scalable enough to integrate new technologies – like PKI (public key infrastructure), smart cards and biometrics – and it must also be reliable, inspiring confidence among your employees, customers, suppliers and partners. In addition to the myriad technologies, products and services that support authentication, authorization and administration, are steps you need to take to make sure that you can resume business if you're temporarily hacked into non-existence. Business resumption planning is yours – but recovery tools, techniques and services belong to your outsourcing partner.

It's time to realistically assess what you can and should do to satisfy an increasingly complicated suite of security requirements. Unless you're really special and very lucky, it's time to call in the cavalry. Find a solid security solutions vendor and stop worrying. The phrase "stay with your core competencies" didn't semantically infiltrate the business lexicon because it was pretty. It's there because it's meaningful.

Technology utilities. Just the sound of the words evokes some weird images. Are technology utilities like gas and electric utilities? Are they real? Will they ever actually work? Lots of very smart people think that within five to ten years we'll all buy computing and communications technology from utilities. If this happens – and I believe it will – then everyone will be outsourcing just about everything (but not strategy). The impact of this will be profound. You will no longer have to deal with any aspect of technology. Do you believe this? Do you even want to believe it? I realize that it's a radical idea, but the major technology vendors have been chipping away at the utility idea for about a decade. Some of the newer technologies – like grid computing – are enabling cornerstones of utility-based services.

If technology utilities emerge, this whole discussion about the acquisition of computing and communications technology is moot. It means that if you're arguing about what to in-source, co-source and outsource, you're missing the point, because you'll eventually outsource everything. OK, we can argue about when this will actually happen, but if it makes any sense to you at all, then the outsourcing you do today is the warm-up for tomorrow. What is it that they say? Practice, practice, practice.

There's one more acquisition strategy we need to talk about: **timing**. Capital markets swing back and forth, sector by sector all the time. The bull market of the 1990s made lots of people who sold technology very rich. Everyone wanted more. First, we had to get PCs and laptops to everyone (even people who didn't work for us). Then we had to make sure our machines kept working after 2000. Then we had to do some e-business. Then we had to do some customer relationship management (CRM), as though we hadn't done any before. All of this added up to what we describe with 20/20 hindsight as the technology bubble. If you bought lots of technology in those days (from companies whose business models weren't subsidized by private equity dollars shoveled out by venture capitalists) you overpaid for technology products and services. But you didn't care because you were making money, because there was a bloated premium in the value of **your** products and services. It was beautiful: everyone was living large on virtual money. But that was then. In the second half of 2000, the capital technology markets collapsed, and the business technology vendors got pummeled by significant spending reductions which, predictably, cratered the stocks of the public technology companies and reduced the valuations of just about every technology company on the planet.

So what does all this have to do with acquisition strategy? Well, if you invested in enterprise data base management, ERP or CRM applications in the 1990s through 2000, you overpaid – but if you bought these applications after the crash you could have used jokes, songs (no matter badly you sing) and wampum for currency. Every December in down capital markets there's an uber-sale on everything (because vendors need to hit their numbers for the year). Outsourcing deals signed before 2000 have been renegotiated. Hourly rates for consultants have fallen so dramatically that some commercial vendors can actually beat in-house rates, though doing so costs them lots and lots of margin. A few years ago Hewlett-Packard offered to buy PriceWaterhouse Consulting (PWC) for $18B. IBM bought the 30,000 consultants in 2002 for $3.5B. At the time of the sale, PWCs projected 2002 revenue was $4.9B (they had revenues of well over $5B in 2001). Based on these rough numbers, IBM stole the company. PWC only received .75 of their expected revenue – an incredibly low valuation, even for the services sector (ironically, the sale may have "saved" PWC from a very rough Initial Public Offering [IPO] which they had filed – and then withdrew when IBM acquired them). Why pick on PWC? It's not PWC I'm picking on. I'm calling attention to capital market swings and how they can be exploited by savvy business technology acquisition managers. If you wanted to hire PWC – or lots of other vendors – during this time, you could have negotiated a pretty good deal.

The stock market versus real estate debate is a great example of how this works. In 2002 everyone was pouring money into real estate, which of course dramatically drove prices up. The equities market was setting five-year and in some cases all time low records. If "buy low/sell high" makes sense to you, then you should have been selling real estate, buying equities – **and hardware, software and services**.

Make sure you track capital market trends and how they impact the pricing of all of your suppliers and vendors. It's amazing how cheap you can buy this stuff when sellers are hungry.

Who Gets The Check?

How much are you spending on technology annually? How does it break down? Are your hardware expenses rising faster than your software expenses? Are your personnel costs rising faster than your hardware and software costs?

Figure 33 converts it all into a matrix that can be used to determine where you are today and where you need to go. Note that the governance issue is absolutely critical to success. If governance – defined here as what is to be done and who is to do it – is mishandled, then the whole business technology products and services acquisition and funding process will collapse.

Figure 34 can help with a current baseline assessment and with the prioritization of funding requirements. This model assumes that the enterprise will "own" the communications networks, data centers, overall security, and the hardware and personal productivity software that runs on this infrastructure – but not the business applications that define the lines of business.

You need to know where you spend your money and how it gets spent. You need to know what gets spent in-house and what gets spent on external consultants.

Figure 34. A funding requirements and planning matrix

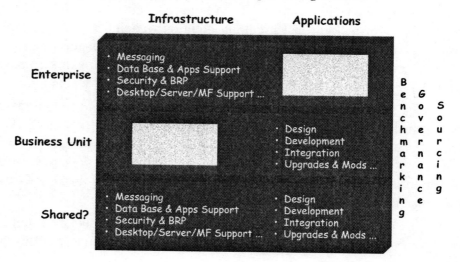

Here's a list to get you started:

Infrastructure

- Messaging Environment (E-mail, Workflow, Collaboration)
- Network and Communication Architecture Design
- Network and Systems Management
- Voice and Data Services & Support
- Infrastructure Engineering Services and Support
- Data Center Hardware (Mainframes, Servers, Switches, Routers, etc.)
- Desktop Hardware
- Laptop Hardware
- Other Access Devices (Personal Digital Assistants, Hybrids)

Applications

- Applications Software Packages
- Applications Development (Integration and Deployment)
- Applications Support

- Applications Architecture Design
- Applications Architecture Services and Support
- Modernization and Migration

Overhead

- Staff (Human Resources, Benefits, etc.)
- Training Services and Support
- General Support (Administrative Assistants, etc.)

You also need to know who "owns" each category and where "collaborative" funding decision-making occurs, should occur and should not occur, given your organizational structure (another good time to think about organizational structures). While this list is helpful, it's only the start. Table 1 suggests that you need to know how much spending occurs and who owns what.

The data is drawn from an informal sampling of over 25 large (Fortune 1000) companies who have decentralized technology organizations. First, it tells us where the money goes – relatively speaking. It also illustrates where the primary responsibilities lie in a decentralized organization. The question

Table 1. Funding practices in a decentralized organization

Categories	Cost	Responsibility Enterprise	Division/LOB
Infrastructure			
Messaging	$$	✓	
Communications/ Network Design	$$$	✓	✓
Network & Systems Management	$$$$	✓	
Infrastructure Support	$$$$	✓	
Hardware Procurement	$$$	✓	

Table 1. Funding practices in a decentralized organization (continued)

Categories	Cost	Responsibility Enterprise	Division/LOB
Applications			
Software Packages	$$	✓	
Development/Integration	$$$$$		✓
Support	$$$$	✓	✓
Architecture Design	$$	✓	
Architecture Services & Support	$$	✓	✓
Modernization & Migration	$$		✓

Categories	Cost	Responsibility Enterprise	Division/LOB
Overhead			
Staff (HR, Benefits ...)	$$	✓	
Training	$$$$$	✓	✓
General Support	$$$$	✓	✓
Research & Development	$$	✓	

remains, however, which is best? Every time two checks appear together, there's a potential problem – which is why governance becomes so important.

Table 1 is also strategic. Look where the money goes. If you spend tons of cash on network and systems management and overall infrastructure support, then you know where to look for core competency assessment, total-cost-of-ownership (TCO) and return-on-investment (ROI) data, the strengths and weaknesses of business cases, and sourcing options. Actually, since nothing's really "cheap," there's potential leverage all over the place.

The applications picture is even more interesting – if for no other reason than the sinkhole called applications development/integration. It doesn't get more expensive than this. If you're hypnotized some time and decide to dramatically customize a packaged application, then you'll discover just how much money it takes to get that extra 5% of functionality out of the application, you know, the 5% you just couldn't live without (the 5% that will cost you the additional 90%).

The overhead numbers are important because we have two categories – training and general support – that are important and often shared among the enterprise and the lines of business. As everyone at this conversation knows all too well, when things are expensive and shared, there will be problems.

Table 1 can also be treated as a test. Try taking it. See what you learn. Do you know the relative costs of your infrastructure, applications and overhead activities? Do you know – unambiguously – where the responsibility lines are drawn?

Governance (see below) refers to two funding dimensions: **how** you fund what you buy and **who** funds what. There are a number of ways to fund business technology products and services in a decentralized organization. Why do I keep talking about decentralized organizations? Because most organizations are decentralized to some extent or another and because distributed computing (client/server, the Internet and the World Wide Web) continues to pull organizations in different directions all the time. Governance makes decentralized organizations work, when it's strong, or fail, when it's weak. The conversation in Chapter V about organizational structures is again relevant! In both proposed structures there's a strong Process Officer with the authority to define and apply governance.

Cost allocation is best applied to infrastructure investments. In the decentralized organization, the enterprise – along with the lines of business – determine what the infrastructure should look like and the likely costs. The enterprise then builds the infrastructure and allocates the costs across the lines of business. Unless you want to get assassinated, allocate the costs according to usage – not equally independent of use. The problem with usage allocation, of course, is the necessary specificity about usage you must produce: you must be able to empirically "prove" that line of business "A" is using more than line of business "B" and should therefore pay more of the infrastructure allocation. Don't get cute with usage metrics. Allocate infrastructure costs according to simple ones, like the number of network connections, amount of data storage, etc.

Fee-for-Service is challenging since in order for it to work the lines of business should be able to look outside for the same services that a central/enterprise group might provide. In other words, fee-for-service works best when free market principles prevail. The obvious problem of course is that the enterprise group often feels like they're at a competitive disadvantage to outside vendors who can use a variety of tricks to win the business (such as fixed price/fixed schedule, loss-leader tactics, and other "best-and-final" techniques). At the same time, you might say that if the internal technology group cannot compete with outside vendors then maybe it's time to outsource the services – and you'd have a good point. Many of us see the fee-for-service model as a mechanism to keep the internal technology groups honest.

Taxation works when there are activities that fall outside of infrastructure and application costs and when there is value attached to these activities. For example, an enterprise technology group might perform analyses of technology or industry trends and make them available to the lines of business. A skunk works might be developed that will test new hardware and software and then share the results with the technologists in the business units. Training often falls under this umbrella as well. The general administrative support of the enterprise group usually falls into this group as does the salaries of technology management. Finally, research and development is often taxable. Where's that Process Officer?

This flip-side of the governance question haunts just about every company on the planet. Basically, the problem looks like this. Organizations want to make the right decisions about collaboration, hardware, software and communications, but are afraid to actually to do so. Decentralized business technology organizations that have weak decision-making governance live in a constant state of fear, uncertainty and doubt. No one ever knows who will actually pay for what, when or how. Strongly governed organizations publish the rules and stick by them. It's all very, very simple and very, very complicated. As always, leadership separates the organized and efficient from the chaotic and wasteful. So what kind of organization do you live in? Here's a short yes-or-no test:

- Do you have to look pretty hard to find benchmarking data?
- Is your organization subjected to religious wars about operating systems, databases, and other standards issues?

- Are the religious wars allowed to continue indefinitely?
- Does everyone have veto power over everyone else, even idiots?
- Can infrastructure investments be made by the lines of business?
- Can applications investments be made by the enterprise infrastructure group?
- Do you spend hours counting MIPS?
- When a debate breaks out is it settled by junior people?
- Do the senior enterprise business managers talk tough, but act like paper tigers?
- Are allocation and taxation funding mechanisms the object of ridicule in your organization?

OK, you guessed it: more than five "Yes's" makes your organization incompetent and silly; eight or more makes it certifiable.

Table 2. Funding governance recommendations

Categories	Cost	Responsibility Enterprise	Division/LOB
Infrastructure			
Messaging	Allocation	✓	
Communications/ Network Design	Taxation	✓	✓
Network & Systems Management	Allocation	✓	
Infrastructure Support	Allocation	✓	
Hardware Procurement	Fee	✓	

Table 2. Funding governance recommendations (continued)

Categories	Method	Responsibility Enterprise	Responsibility Division/LOB
Applications			
Software Packages	Fee		✓
Development/Integration	Fee		✓
Support	Fee		✓
Architecture Design	Tax	✓	
Architecture Services & Support	Fee		✓
Modernization & Migration	Fee		✓

Categories	Cost	Responsibility Enterprise	Responsibility Division/LOB
Overhead			
Staff (HR, Benefits ...)	Taxation	✓	
Training	Taxation	✓	✓
General Support	Taxation	✓	✓
Research & Development	Taxation	✓	

Table 2 suggests what the landscape should look like. It identifies the funding mechanisms and responsibilities that make the most sense in decentralized organizations.

Take a look at your organization with reference to this table. How does it look?

As Table 2 suggests, **infrastructure funding** is largely the responsibility of the enterprise. Infrastructure design and construction is funded via taxation and allocation. The design of the infrastructure benefits everyone, so it's taxable. But **use** of the infrastructure will vary from business unit to business unit and should therefore be funded by usage allocation.

The arguments to watch out for include:

- Business unit "A" is deploying a lot more e-business applications than "we" (business unit "B") are ... so why should we pay for distributed security at the same rate as them? Maybe the costs should be spread across taxation, fee-for-service and allocation funding mechanisms.
- I am not willing to pay for remote access infrastructure. I don't use it today and don't expect to use it tomorrow ... make business unit "Y" pay for it ... they're the ones sending everyone home to work!
- Tell me again why I must be taxed at the same rate as business unit "P" that requires ten times the training I require?

You get the idea. The key to the successful implementation of mixed funding mechanisms is – you guessed it – serious leadership that expresses itself in a well-defined governance policy. If arguments like the ones listed above are allowed to infect your organization, you'll be spending as much time on resolving arguments as you will on building and supporting your infrastructure and applications.

The simple rules should be:

- **Allocation** = If you use it you pay for it.
- **Taxation** = There are some things good for everyone and therefore everyone's going to pay for them – no questions, no debates and no returns.
- **Fee-for-Service** = There are IT decisions that the lines of business can make, decisions about applications, about desktop upgrades, and the like that they can certainly make on their own ... and they have the right to look internally or externally to implement those decisions ... if they decide they want, they pay for it.

Is all this simple? No. Here's why. Much of what we're talking about in what appear to be clearly defined terms are in practice ambiguous and fuzzy. For example, when does the purchase of new laptops move from fee-for-service to allocation? If the central infrastructure organization runs the data center that houses mostly mainframe applications, but not some of the important distributed applications that reside on servers in the lines of business, where should the line be drawn between fee-for-service, allocation and even taxation, if either the mainframe or distributed applications team used a common applications architecture (developed by the enterprise group via taxes)?

But what if you're organization is centralized? Does all this change? You bet. Pendulums swing back and forth. During the 1990s, the trend was toward decentralization where divisions and lines of business were given lots of autonomy. Some companies even allowed their parts to run their own infrastructures. By 2000, lots of these companies were re-thinking these relationships and began to re-centralize. What makes sense? It obviously depends (what a lame answer, huh?). But it really does "depend." What's unambiguous? Infrastructure centralization is unambiguous. Regardless of how decentralized your applications environment is, it always makes sense to centralize your computing and communications infra-structure. The open questions pertain to applications development, integration and deployment. Who should own the applications? And the data that enables them? Collaboration requires lots of centralization. More specifically, collaboration requires the centralization of the computing and communications infrastructure, data architectures and application architectures. This means that while lines of business can select (or develop, if they're nuts) the applications they need to compete, they must do so within some strict architectural parameters that will make the applications run efficiently on the company's shared computing, communications and data infrastructure. As we discussed in Chapter V, there are fewer degrees of freedom here than met the eye – or the expectations of cowboys in the lines of business who are always poised to secede from the technology union. It's possible to share power across the enterprise and the lines of business but only when there's strong governance in the company about where the lines get drawn. If they're fuzzy and senior management is unwilling, when necessary, to flex its muscles, then the path to collaboration through integration will always be bumpy. Hell, it won't just be bumpy – it will be mined!

What about sourcing? If the lines of business decide that they'd like to shop a bid for services internally **and** externally and select an external vendor, who manages that vendor if the vendor's work requires them (as it will inevitably) to interface with existing internal policies, procedures, hardware and software? If the enterprise organization hires an outside vendor to help it perform infrastructure support for the lines of business, does the outsourcer report to the enterprise group or the line of business?

Disputes should be handled via some form of published grievance procedure. If a line of business feels it has a legitimate gripe, there should be a process that helps resolve the dispute. If you want the arbitration process to work you will have to use external judges.

Applications are the lifeblood of the lines of business – and they should pay the freight here as well (as suggested in Table 3). But there are some major issues surrounding applications funding that you should be aware of as you develop a funding strategy. Here are several of the most important ones:

- Applications that were initially intended to support local users who become remote users will have to be re-engineered. If remote access performance is sub-standard, it's often assumed that the infrastructure is to blame when more often than not it's the application.

- Applications should be reviewed by the infrastructure support team before they're built to make sure that support requirements are not prohibitively expensive. End-to-end applications planning should become a best practice.

- Applications integration often involves integration to back-end legacy databases, databases that are often maintained by enterprise database administrators. Make sure that the infrastructure support team is synchronized with applications integration efforts.

- Web-to-legacy connectivity is a common way to rapidly deploy e-business applications. But, in order to design, develop, deploy and support such applications both infrastructure and applications professionals will be necessary. Make sure they are well coordinated.

The most cost-effective way to develop applications is via an enterprise applications architecture (the specification of platforms, tools, development environments, databases, browsers, and the like). If your organization does not have a common architecture then you'll re-invent the wheel over and over again. The cost of a single architectural specification will be much, much lower than the costs of repetition. Of course, without a governance policy that sees that the architecture gets used (and re-used), the investment will be wasted.

Many Happy – And Miserable – Returns

Let's keep return-on-investment (ROI) and total-cost-of-ownership (TCO) in perspective. You cannot build a business with these hammers. Are they important? Yes. Insightful? Yes. Strategically and tactically useful? Yes. But obsessive-compulsive TCO/ROI behavior is as unhealthy as any obsessive-compulsive behavior.

The ultimate argument for a business technology initiative is made in the business case (which we'll talk about momentarily). But key to good business cases is qualitative and quantitative data about the cost of the project's entire life cycle and the strategic impact it will have on collaborative business processes.

Let's start with costs. There are acquisition costs, operational costs and softer costs that are by nature more difficult to quantify. The total-cost-of-ownership (TCO) also includes costs over the entire life cycle of the hardware or software product. Softer costs include the cost of downtime, internal consulting that comes from indirect sources, and costs connected with your degree of standardization. There's the cost of developing an application and the cost of supporting it over the course of its life: did you know that it costs roughly five times as much to support an application as it does to develop it?

TCO data is an essential part of your overall business case which should ultimately be driven by the strategic and/or tactical return you expect to get from your investment. In other words, TCO data drives return-on-investment (ROI) data – which, like TCO data – is both hard and soft. Herein lies the controversy about ROI calculations.

Is it important to ask meaningful questions about why a business technology initiative exists? And what impact it will have on business (if it goes well)? Of course. So why is there so much disagreement about ROI? An excellent **CIO Insight** Research Study reported that while lots of ROI methods are used, by far the most popular were ones that calculated cost reduction, customer satisfaction, productivity improvement and contributions to profits and earnings.[1] Two years is also considered by the majority of business technology executives as a reasonable time over which to measure ROI.

So What Are The Methods?

One of the easiest is based on a simple calculation that starts with the amount of money you're spending on a business technology initiative and then calculates the increased revenue or reduced costs that the investment actually generates. If a project costs a million bucks but saves two million then the ROI is 100%. Not too bad, and a good ROI method to implement as a first step.

Another simple one is based on payback data – the time it takes to offset the investment of the business technology initiative through increased revenues or reduced costs. If the payback period is short – and the offsets are great – then the project is "successful."

There are also methods based on economic value analysis or value added (EVA), internal rates of return (IRR), net present value (NPV), total economic impact (TEI), rapid economic justification (REJ), information economics (IE) and real options valuation (ROV), among lots of others.[2]

What about soft ROI? In the mid- to late-1990s, companies developed Web sites for a variety of reasons. First generation sites were essentially brochureware, where very few transactions took place. What was the ROI on these sites? They did not reduce costs: in fact, they increased them. Nor did they generate revenue. They were built to convince customers, Wall Street analysts, investors and even their own employees that they "got it," that they understood that the Web was important. An intangible benefit? Absolutely.

Everything in this business technology conversation is about mindset and process. The same is true of TCO and ROI calculations. While I think that anyone who launches a business technology project without TCO and ROI data is nuts, I can also appreciate the need for balance and reasonableness. This is why there's so much controversy about TCO/ROI. Lots of senior people think that too rigorous of an application of TCO and ROI methodology will distort projects and perhaps even undermine business results. Others think that the effort to collect and analyze TCO and ROI data is disproportionate to the returns. What to do?

Simplicity – as usual – is our friend.

Adopt a flexible approach to TCO and ROI. TCO data should feed ROI data which should feed the overall business cases for business technology decisions. Hard data is always better than soft data, but soft data – if it can be monetized somehow (like generating a premium for your stock price or enhancing your brand) – should also be analyzed.

The simplest approach to TCO data collection and assessment is a template that requires the collection of specific hard and soft data, and the simplest approach to ROI data collection and assessment is based on simple metrics that measure payback over a reasonable period of time. Payback should be defined around internal metrics – like cost reduction – and external ones, like improved customer service and profitability.

TCO and ROI should not be used as clubs to hit people over the head with: they should be used to inform decisions and monitor progress. They should also play a role in the death of projects-gone-berserk.

The Business Of Business Technology Cases

How do you sell a technology project to a battered senior executive team that's in no mood to spend a ton of cash on a new project, especially after you've educated them about the 25% chance of success?

Projects come from lots of different places. Cocktail parties, directors, in-flight magazines, ball games, dinner parties, Gartner reports, advisors, spouses, loss of market share, falling margins, arguments and security breaches, among other predictable and unpredictable sources. Every single project should be based on a rigorous analysis of a solid business case. Every single project should be conceived and evaluated with three possible outcome decisions: go/no go/we need more information. All three should have an equal chance of winning.

Business case development is all about the identification of real and political reasons to buy something or engage a consultant. This can be a very tricky process, since (because of the perennial competition for funds) there will always be project enemies, those just waiting to say, "I told you so" when the project goes south. Technology buyers - especially in large enterprises - have to make sure they've covered their flanks. **The "business case" is therefore as much a "real" document as it is a political one.** When you read – or write – them, make sure you use at least these two lenses.

Business cases are typically organized around questions designed to determine if the investment makes sense. Let's look at the key questions that you should answer before buying hardware, software, communications, or consulting services. Let me say up front that the answers to these questions should be in a ten-page document (with appendixes if you must) with a one page executive summary. If you generate a long treatise on every investment, you'll never get the time or respect of busy senior executives. Would you read a 50-page business case?

The first step is to identify – and answer – collaborative business value questions or, put another way, answer one simple question: how will this project help the business collaborate profitably?

Here are the questions:

Key Collaborative Business Questions

What Collaborative Business Processes are Impacted by the Investment/Application(s)?

The correct answer here identifies a process that's broken, a process for which you have empirical benchmarking data. For example, if you were serving customers with call centers, you'd have to know what you're spending now and that the costs were rising dramatically, or customer satisfaction levels were falling. You'd then need to know exactly what your performance target should be (e.g., reduce costs by 25%, increase satisfaction by 25%).

How Pervasive is the Product or Service among your Traditional and Unconventional Competitors?

Many decisions about business technology adoption are driven by what the competition is doing. It's always easier to sell something in-house when your competitors have already adopted the product or service (of course, this assumes that your competitors are smart). Be prepared for the contrarian argument here: "if those people are doing it, we sure as hell shouldn't be!"

What Competitive Advantages Does/Will the Product or Service Yield?

In other words, why is this product or service so great and how will it help you profitably grow the business?

How Does the New Product or Service interFace with Existing Collaborative Business Models and Processes?

This is a big question, since the "wrong" answer (like: "it doesn't") will kill a project. If deployment means ripping out or enhancing infrastructure, then you've just significantly increased the price (and given your internal enemies ammunition). You've also raised a legitimate yellow flag.

Key Technology Questions

How Mature is the Product or Service?

The point of these questions is to determine if the technology or service actually works – and works well. You will need quantitative data here: it won't help to tell everyone about how great your brother-in-law's company is. Additional questions here concern scalability, security, modifiability, usability, etc.

What Share of the Market Does the Product or Service Own?

If the answer to the question is: "well, 1%," then a major explanation will be necessary. Remember to stay mainstream, anyway. Why do you want to be an early adopter of someone's half-baked product?

How Does the New Product or Service Integrate with the Existing or Planned Communications and Computing Infrastructure?

This question is of course the second most important question you need to ask – and answer – because it identifies any problems downstream that might occur because of decisions already made (that cannot be undone, or would be prohibitively expensive to undo).

Key Cost/Benefit Questions

What are the Acquisition, Implementation and Support Costs and Benefits?

Here you need to look at obvious costs, like software licenses, and less obvious ones, like training, indirect user and help desk support, as well as the expected operational and strategic benefits like expense reduction, increased market share, improved customer service, increased cross- and up-selling, improved customer retention, etc. Here's where total-cost-of-ownership (TCO) data gets inserted.

What are the Support Implications? How Complex, How Costly, How Extensive, How Timely?

Support is extremely important. You need to know - empirically - what the support requirements and costs will be defined in terms of $$$, people and time.

What are the Migration Issues? How Complex, How Costly, How Extensive, How Timely?

This may or not be relevant, but if another tool is in place, you have to answer questions about how to get from one to the other.

Key Risk Questions

What are the Technical, Personnel, Organizational, Schedule and other Risks to be Considered? What's the Risk Mitigation Plan?

The risk factors that everyone will worry about include scope creep, cost and time overruns, incompetent or irritable people, implementation problems, support inadequacies, training problems, and the like.

Copyright © 2005, Idea Group Inc. Copying or distributing in print or electronic forms without written permission of Idea Group Inc. is prohibited.

If risks are assessed as medium or high then a mitigation plan must be developed, a plan that either moves the risk down to the "low" category or eliminates it altogether. If the risks remain high, the project sale is dead.

The Recommendation

The recommendation is **go/no go/we need more information**. The whole purpose of integrating a business case into the sales pilot is to avoid the "we-still-need-more-information-syndrome."

The business case should also identify at least two "accountable" people, people whose reputations rest to some extent on the success of the project. One should be from the technology side of the organization and one from the business side (if you can find the right hybrid). If there are no project champions, it's time to go home.

Who owns all this? If there's no Process Officer in your company, no anointed process, and no business case requirement, you're in trouble. It's amazing how crazy all this is. Business cases are not that difficult to create. What's difficult is swallowing them. Cultures that are especially political have the worst digestion problems – and make the most business technology investment mistakes.

Project Management Redux

This one's been around for a long, long time. It's actually masqueraded under lots of aliases over the years including total quality management (TQM), capability maturity, statistical quality control and balanced scorecards, among other attempts to better organize us at work. Of course all of these movements have their gurus and disciples, and all of them are different in their own important ways, but by and large they're about professional discipline – which, of course – very few of us actually have – which is why these movements come and go.

But project management **is** different because it's never really achieved star status. It just keeps on rolling along. Maybe that's why it never makes the top five initiatives-we-need-to-worry-about-this-month, but **always** makes the top ten we're worrying about this year. It's a real discipline that we don't practice very well, but always need to improve.

At the outset of this discussion, let's stipulate that discipline – like adhering to technology standards or hiring only smart, hard-working, ethical people – is hard, primarily because it means that we have to control our impulses. And we hate that. But discipline works: it makes money. And project management should be a core discipline in every company in the world.

So what's the problem? The first thing we need to do is declare project management an important thing. Then we need declare this very publicly. If you're in a big company, then you need to organize a serious off-site to raise consciousness. If you're a medium-sized company then you need to speak to all of the employees and incentivize the change agents. But the real key to this is consistency and persistence. Initiatives fail because companies roll out programs, processes and policies that vary from group to group or organization to organization, or because they lose interest over time, something that employees can smell long before the plug is officially pulled. I remember well sitting in audiences listening to senior executives talking about the company's major new initiative only to hear the lifers muttering "this too shall pass." The "been there/done that" problem is a big one, especially if your track record is weak. Lots of long-term employees have adapted to management's lukewarm commitment to major initiatives and have learned to drag their participatory feet for as long as they can (or until management loses steam).

Enough politics. What should your project management expertise look like? What skills do you need – really? Here they are:

The ability to assess a project's likely success or failure. Note that this assessment comes after the business case has been filed – and approved. Can the project be successful? What are the risks? Who's the best person to lead the project? Is a good team available? What are the immediate problems we have to solve?

The ability to keep the business case front-and-center as the project unfolds. Once approved, the business case is the blueprint developed by well-dressed, well-meaning architects. The project is all about construction – schedules, sub-contractors, screw-ups, miscommunication, etc. You know, reality. Companies need to continuously link project progress to the business case that gave birth to the project in the first place. This means that the business case also needs to be re-assessed on a regular basis, especially if the project is a big, long one. Measure the distance between the business case and project progress on a regular basis. If they start to drift away from each other, then it's time to take action. (This measure, by the way, is seldom used, which is why projects begin to take on lives of their own.)

The ability to execute project fundamentals, such as milestones, deliverables, schedules, cost management, reviews, etc. These skills are not necessarily resident in your company. You might consider getting a number of your good project managers certified in the latest thinking, processes and tools. The Project Management Institute (www.pmi.org) is a good place to start.

The ability to kill bad projects and the ability to determine if a project is hopeless or salvageable. How do you kill projects? This is like: "should I sell that stock that's down 50% from when I bought it"? Or, can this project come back from the dead (this stock is only hibernating – it will be back)? There are at least three reasons to kill a project:

- The business case and project are drifting far apart. The business assumptions about the importance of the project are no longer valid.
- Project execution is way off track; the project is over-budget, behind schedule, etc.
- The probability of recovery is low.

Let's talk about these areas.

First, business case ←→ project distance needs to be measured at least quarterly – in the case of small projects – and monthly, in the case of big ones. The distance is measured in terms of the project's expected (strategic or tactical) impact: new metrics – PESIs and PETIs (we could probably organize a conference around these two). This distance is essential to project survival: if it appears that the strategic or tactical impact of a successful project will be minimal, then kill the project. How do you know if it's likely to be minimal? If the requirements that supported the business case have changed, if the competitive landscape has changed (rendering the project "obsolete"), or if the assumptions about cost prove invalid.

Project execution is more quantitatively measured. Here all you have to do is measure estimated versus actual project performance. If the schedule, milestones, costs, deliverables, and risks are 33% or more out of sync with your estimates, then the project is out of control and unlikely to recover. If two or three of the indicators are 20% to 25%, then the project should be flashing yellow and tracked closely to see if it goes red.

Finally, if a project is drifting from its strategic or tactical objectives, or if execution is poor, a judgment must be made about the likelihood of turning the project around. Like the stock you bought that tanked, can the project come back? The hardest decision to make is one that turns the lights off for good. But if the business case ←→ project distance is great – and growing – and the execution is poor – and getting worse – it may be time to pull the plug. Recovery is unlikely when both trajectories are in the wrong direction.

How do you "see" all of your projects? Some companies have weekly project meetings, some have monthly ones, and politically anal companies only do it informally, privately. The time and effort necessary to track multiple projects – if the number of projects is high enough – will redefine your job as a full-time project manager. This is a bad thing. What you need is a **dashboard** that immediately shows you which projects are on track and which are under-performing – and which are candidates for capital punishment.

Dashboards are not hard to build. They're even easier to buy. Microsoft Project can be used to feed off-the-shelf reporting applications or you can customize one to show projects as red, yellow or green, as well as the trends. You should standardize on both the project management and dashboard application. You should also make sure that accurate information gets into the dashboard, which should run on your desktop, laptop and PDA. In other words, it should be possible to check on major projects anytime you want. Some of these tools – like The Project Control Panel, developed by the Software Program Managers Network (www.spmn.com/pcpanel.html) – extract data from Microsoft Project and inject it into a Microsoft Excel tool that displays project status. Other tools, like Portfolio Edge from Pacific Edge Software, enable you to track multiple projects at the same time.

You need to make the rules. For example, you might have a 10% to 15% estimated/actual variance rule that triggers weekly project meetings. You might have a rule that says that variation on project deliverables is more important than schedule variation, and you might have one that triggers some project survival rules. The key is to field a set of rules that work for your company, your culture and project management experience.

Finally, groups of projects = programs and all of your programs = your portfolio. Strategists manage portfolios, tacticians manage projects. It's important to roll your projects up into programs and portfolios to keep the big picture in perspective.

Governance

Governance is real, unavoidable and political. Increasingly, it's stimulated by regulatory changes. To some, it defines corporate culture. Others only think about governance as applicable to back-office processes. The approach to governance I'm taking here is broader than most definitions you'll find. In fact, I'm actually making an argument for the concept of governance to change dramatically, for us to treat governance as a way to manage internal **and** external priorities through the application of a set of governance best practices.

Ultimately, governance is about things to do (and not do) – the "rules" – and the policies and procedures around how we adhere to governance – and especially what happens when we don't. Governance is inescapably tied to leadership. We like to think in terms of governance "styles" that – if the truth be told – refer to the location and use of power. Styles can be open, closed, participatory, democratic, federated or monarchical. Ultimately, these styles say a lot about cultures and our ability to govern ourselves – and co-govern with our suppliers, partners and even customers.

All of the business books and articles published each month about leadership, innovation, customer relationship management, and e-business – among hundreds of other topics – are as much about governance as anything else. Does your company have a Web site? Of course. Is there a standard applications architecture for the site, guidelines for all of your business units to follow about how to design and develop their sites? No? Then you have weak technology governance. Who owns strategy in the company? Is strategic planning ad hoc? Or is it structured and scheduled? Structured and scheduled strategic planning happens in companies that enjoy clarity about who does what, when and how, in companies with strong governance.

Who responds to government regulations? Compliance requirements? The Healthcare Insurance Portability and Accountability Act (HIPAA) is real. Sarbanes-Oxley is real. Regulation full disclosure (Regulation FD) is real. Who keeps your company compliant? It's all about governance.

What about organization responsibility and authority? When the two are distinct, there's usually a governance problem. When they are intertwined, efficiencies are possible. We also tend to link governance to controversy when in fact governance – like simplicity – is our friend. But as friends go, remember that governance demands loyalty and consistency.

This section looks at governance as it occurs within and beyond internal corporate functions and activities. It suggests a new way to think about governance. But the key to the approach taken here is its anchor in business technology best practices regardless of whether you're developing a new business strategy or deploying a new local area network. This is consistent with the view that business, technology and management best practices are inextricably entwined, that it's impossible to talk about technology governance without also talking about business and management governance.

Let's first identify all of the activities that need to be governed. We'll use five business technology "layers" to do this. Then we'll look at the procedural-regulatory context in which all of these activities occur, then at alternative organizational structures, and then we'll assign governance roles to each activity within each organizational structure that I discuss. The long and short of it is that once you understand how you're functionally organized, and once you understand all of the things that occur within those functions, once you understand the policies, procedures and regulations that need to be governed

Figure 35. The range of governance

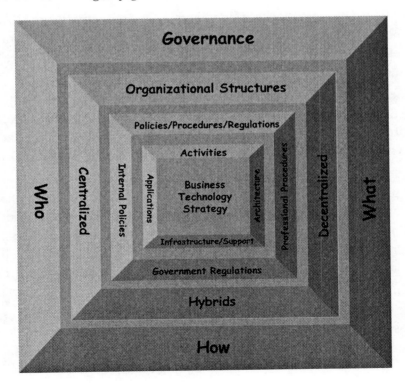

– and once you understand the organizational structures that manage governance - it's pretty easy to identify the governance issues and responsibilities. The reality is that organizational structure tends to bifurcate into flavors of "centralized" and "de-centralized," usually when there are quasi-independent business units within a company. Ah, but here's the rub: everything turns on the definition of "quasi" and "independent" – and on the desirability of having quasi-independent business units (however the boundaries are defined). These hugely important distinctions – as well as an approach to process governance that might work for you – are discussed here.

Governance is the administrative platform on which everything sits, in which everything occurs. We've talked about functions, activities, policies, procedures, regulations and organizational structures. It's now time to turn to the governance activity itself and suggest who should govern what, as Figure 35 suggests. I'll end this section with a look at exactly how all this can be accomplished.

Activities occur within organizations, and organizations can be defined around their cultures along some centralization – decentralization continuum. The governance agenda should be defined around specific activities and interpreted through your culture and, ultimately, your organizational structure.

I've organized the governance assignments around the activities and structures in a set of matrices. The purpose of the matrices is to identify the range of governance assignments and to facilitate a discussion about what makes sense and what doesn't – and why.

Let's start with the governance of specific activities in a centralized enterprise. Figure 36 begins the analysis with a look at the activities across the five layers already discussed. Governance is indicated in the figure by the large and small checks in the columns. A large check indicates strong governance authority, while a small one indicates weak governance authority.

Figure 36 suggests that all of the activities in a centralized organization should be governed by the centralized enterprise. But you'll notice that there's an additional column in the matrix that refers to suppliers, partner and customers. Another way of understanding the additional column is to expand the definition of governance to include activities within and beyond the corporate firewalls.

This is an important recommendation, since as we expand our transactions to include suppliers, partners and customers, the whole concept of governance changes. Take, for example, governance around technology standards. There are internal governance issues, such as the standardization of machines, but

Figure 36. Activities governance matrix for centralized organizations

Activity (By Layer)	Centralized Enterprise	Suppliers, Partners & Customers
Business Technology Strategy Layer **Collaboration** – activities that focus on the inter-connected marketplace and inter-networked companies; steps to get customers, employees, partners and suppliers to inter-connect; activities to make collaboration work across application, data and communications architectures ...	✓	✓
Customization & Personalization – activities and projects that focus on mass personalization, behavioral models to correlate online and offline behaviors, wireless personalization and personal and professional customer relationship management (CRM), among other related areas ...	✓	✓
Supply Chain Management – activities that include supply chain concepts, models and tools, as well as the implementation and management of integrated supply chains in the context of SCM standards, technologies (such as exchanges), and the leading SCM platforms ...	✓	✓

Figure 36. Activities governance matrix for centralized organizations (continued)

Activity (By Layer)	Centralized Enterprise	Suppliers, Partners & Customers
Business Technology Strategy Layer **Business ←→ Technology Convergence Strategy** – activities that focus on methods for developing and assessing collaborative/integrated business technology strategies; activities that focus on current business scenario that's driving your collaborative business strategy and the plan for integrating computing and communications technology in your company ...	✓	✓
Competitor Intelligence – work that focuses on specific competitors including information about their sales, marketing, profitability, key employees, strategy, tactics, etc. ...	✓	✓
Business Process Management – activities that focus alternative business process modeling methods, tools and techniques; business processes and models measurement ...	✓	✓

Figure 36. Activities governance matrix for centralized organizations (continued)

Activity (By Layer)	Centralized Enterprise	Suppliers, Partners & Customers
Strategic Business Applications Layer **Application Optimization** – activities that look at major technology and business processes and how they can be optimized with a variety of business applications, like customer relationship management (CRM), enterprise resource planning (ERP), and other applications …	✓	✓
Core Business Applications Management – activities that identify the applications that make money for companies as well as applications that define the company's competitive advantages …	✓	✓
Business Analytics – activities that focus on the processes and technologies that yield insight from sales, marketing, customer service, finance, accounting, technology infrastructure and competitor data …	✓	✓

Figure 36. Activities governance matrix for centralized organizations (continued)

Activity (By Layer)	Centralized Enterprise	Suppliers, Partners & Customers
Enterprise Business Technology Architecture Layer **Applications Architectures** – activities that focus on how mainframe (single tier), client-server (2 tier) and Internet/ Intranet (3 tier → n tier) applications have changed and what the trade-offs among the architectures (defined around flexibility, scalability, reliability, etc.) ...	✓	✓
Communications Architectures – activities that focus on existing and emerging communications networks including especially the role that wireless access and transaction processing will play in emerging collaborative business models and processes ...	✓	✓
Data Architectures – activities about the role that data, information and knowledge play in collaborative transaction processing; work on existing data base management platforms, data warehousing, data mining and knowledge management, especially as they contribute to business intelligence ...	✓	✓

Figure 36. Activities governance matrix for centralized organizations (continued)

Activity (By Layer)	Centralized Enterprise	Suppliers, Partners & Customers
Enterprise Business Technology Architecture Layer **Security Architectures** – activities about security and privacy inside and outside of corporate firewalls; activities about authentication, authorization and administration technologies and tools ...	✓	✓
Business Scenario Development – activities about current and emerging business models and processes and the ability to map them in current and future competitive contexts ...	✓	✓
Enterprise Technology Architecture Modeling – activities about the overall organization of technology that supports overall business goals, especially as all of this integrates and works as seamlessly as possible ...	✓	✓
Enterprise Architecture – activities about the overall business-technology architecture, especially how it'd defined and how it adapts to changes in business and technology; the overall blueprint for business technology optimization ...	✓	✓

Figure 36. Activities governance matrix for centralized organizations (continued)

Activity (By Layer)	Centralized Enterprise	Suppliers, Partners & Customers
Infrastructure Layer **Messaging/Workflow/Calendaring** – activities that deploy the platforms that support all varieties of communication and how communications technology enables communication and transactions among employees, customers and suppliers …	✓	✓
Automation – activities that focus on the applied potential of intelligent systems technology and the application of that technology to personal and professional automated transaction processing, monitoring, e-billing, and the like …	✓	✓
Data Base/Content/Knowledge Management & Analysis – activities that position data, information, knowledge and content – of all varieties (static, dynamic, text, video, etc.) – and how it can be managed for alternative purposes, as well as data, knowledge and content management platforms, next generation data base management applications …	✓	✓

Figure 36. Activities governance matrix for centralized organizations (continued)

Activity (By Layer)	Centralized Enterprise	Suppliers, Partners & Customers
Infrastructure Layer **Integration & Interoperability** – activities that make disparate, incompatible applications, standards, data, platforms and architectures communicate with one another, focusing on enterprise applications integration (EAI) and Internet applications integration (IAI), wrapper/glue technologies like XML, as well as more conventional middleware ...	✓	✓
Support Layer **Desktop/Laptop/PDA Support** – activities that focus on the management of network access devices ...	✓	✓
Data Center Operations – activities about how to organize and manage data centers ...	✓	✓
Server Farm Design & Maintenance – activities that focus on how to design server architectures and support server farms ...	✓	✓

Figure 36. Activities governance matrix for centralized organizations (continued)

Activity (By Layer)	Centralized Enterprise	Suppliers, Partners & Customers
Support Layer **Network Design & Support** – activities that focus on the design and support of local area networks, wide area networks, virtual private networks and the Internet ...	✓	✓
Security & Privacy – activities that focus on security architectures, authentication, authorization, administration and business resumption planning ...	✓	✓
Procurement & Asset Management – knowledge that focuses on how to procure and manage computing and communications assets ...	✓	✓
Business Technology Acquisition Strategy – activities that focus on all aspects of the technology procurement and support process, including especially in-sourcing, co-sourcing and outsourcing ...	✓	✓
RFP & SLA Development – activities that focus on the development of requests for proposals (RFPs) and service level agreements (SLAs) necessary to optimize the business technology sourcing process ...	✓	✓

there are also external standards issues, such as the flavor of extensible markup language (XML) that your industry adopts. Just as suppliers to Wal-Mart have known for years, they have to adhere to the standards set by the members of the Wal-Mart supply chain.

Governance is internally selfish, but shared externally. This sets up a conflict between our instincts to control our internal environment and the need to share control with the members of your extended transaction family.

Figure 37. Activities governance matrix for decentralized organizations

Activity (By Layer)	Enterprise	Business Units	Suppliers, Partners & Customers
Business Technology Strategy Layer **Collaboration** – activities that focus on the inter-connected marketplace and inter-networked companies; steps to get customers, employees, partners and suppliers to inter-connect; activities to make collaboration work across application, data and communications architectures ...	✓	✓	✓
Customization & Personalization – activities and projects that focus on mass personalization, behavioral models to correlate online and offline behaviors, wireless personalization and personal and professional customer relationship management (CRM), among other related areas ...	✓	✓	✓

Figure 37. Activities governance matrix for decentralized organizations (continued)

Activity (By Layer)	Enterprise	Business Units	Suppliers, Partners & Customers
Business Technology Strategy Layer **Supply Chain Management** – activities that include supply chain concepts, models and tools, as well as the implementation and management of integrated supply chains in the context of SCM standards, technologies (such as exchanges), and the leading SCM platforms ...	✓	✓	✓
Business ←→ Technology Convergence Strategy – activities that focus on methods for developing and assessing collaborative/integrated business technology strategies; activities that focus on current business scenario that's driving your collaborative business strategy and the plan for integrating computing and communications technology in your company ...	✓	✓	✓

Figure 37. Activities governance matrix for decentralized organizations (continued)

Activity (By Layer)	Enterprise	Business Units	Suppliers, Partners & Customers
Business Technology Strategy Layer **Competitor Intelligence** – work that focuses on specific competitors including information about their sales, marketing, profitability, key employees, strategy, tactics, etc. ...	✓	✓	✓
Business Process Management – activities that focus alternative business process modeling methods, tools and techniques; business processes and models measurement ...	✓	✓	✓

Figure 37. Activities governance matrix for decentralized organizations (continued)

Activity (By Layer)	Enterprise	Business Units	Suppliers, Partners & Customers
Strategic Business Applications Layer **Application Optimization** – activities that look at major technology and business processes and how they can be optimized with a variety of business applications, like customer relationship management (CRM), enterprise resource planning (ERP), and other applications ...	✓	✓	✓
Core Business Applications Management – activities that identify the applications that make money for companies as well as applications that define the company's competitive advantages ...	✓	✓	✓
Business Analytics – activities that focus on the processes and technologies that yield insight from sales, marketing, customer service, finance, accounting, technology infrastructure and competitor data ...	✓	✓	✓

Figure 37. Activities governance matrix for decentralized organizations (continued)

Activity (By Layer)	Enterprise	Business Units	Suppliers, Partners & Customers
Enterprise Business Technology Architecture Layer **Applications Architectures** – activities that focus on how mainframe (single tier), client-server (2 tier) and Internet/ Intranet (3 tier → n tier) applications have changed and what the trade-offs among the architectures (defined around flexibility, scalability, reliability, etc.) ...	✓	✓	✓
Communications Architectures – activities that focus on existing and emerging communications networks including especially the role that wireless access and transaction processing will play in emerging collaborative business models and processes ...	✓	✓	✓

Figure 37. Activities governance matrix for decentralized organizations (continued)

Activity (By Layer)	Enterprise	Business Units	Suppliers, Partners & Customers
Enterprise Business Technology Architecture Layer **Data Architectures** – activities about the role that data, information and knowledge play in collaborative transaction processing; work on existing data base management platforms, data warehousing, data mining and knowledge management, especially as they contribute to business intelligence …	✓	✓	✓
Security Architectures – activities about security and privacy inside and outside of corporate firewalls; activities about authentication, authorization and administration technologies and tools …	✓	✓	✓

Figure 37. Activities governance matrix for fecentralized organizations (continued)

Activity (By Layer)	Enterprise	Business Units	Suppliers, Partners & Customers
Enterprise Business Technology Architecture Layer **Business Scenario Development** – activities about current and emerging business models and processes and the ability to map them in current and future competitive contexts ...	✓	✓	✓
Enterprise Technology Architecture Modeling – activities about the overall organization of technology that supports overall business goals, especially as all of this integrates and works as seamlessly as possible ...	✓	✓	✓
Enterprise Architecture – activities about the overall business-technology architecture, especially how it'd defined and how it adapts to changes in business and technology; the overall blueprint for business technology optimization ...	✓	✓	✓

Figure 37: Activities Governance Matrix for Decentralized Organizations (continued)

Activity (By Layer)	Enterprise	Business Units	Suppliers, Partners & Customers
Infrastructure Layer **Messaging/Workflow/Calendaring** – activities that deploy the platforms that support all varieties of communication and how communications technology enables communication and transactions among employees, customers and suppliers ...	✓	✓	✓
Automation – activities that focus on the applied potential of intelligent systems technology and the application of that technology to personal and professional automated transaction processing, monitoring, e-billing, and the like ...	✓	✓	✓

Figure 37: Activities Governance Matrix for Decentralized Organizations (continued)

Activity (By Layer)	Enterprise	Business Units	Suppliers, Partners & Customers
Infrastructure Layer **Data Base/Content/ Knowledge Management & Analysis** – activities that position data, information, knowledge and content – of all varieties (static, dynamic, text, video, etc.) – and how it can be managed for alternative purposes, as well as data, knowledge and content management platforms, next generation data base management applications ...	✓	✓	✓
Integration & Interoperability – activities that make disparate, incompatible applications, standards, data, platforms and architectures communicate with one another, focusing on enterprise applications integration (EAI) and Internet applications integration (IAI), wrapper/glue technologies like XML, as well as more conventional middleware ...	✓	✓	✓

208   Andriole

Figure 37. Activities governance matrix for decentralized organizations (continued)

Activity (By Layer)	Enterprise	Business Units	Suppliers, Partners & Customers
Support Layer **Desktop/Laptop/PDA Support** – activities that focus on the management of network access devices ...	✓	✓	✓
Data Center Operations – activities about how to organize and manage data centers ...	✓	✓	✓
Server Farm Design & Maintenance – activities that focus on how to design server architectures and support server farms ...	✓	✓	✓
Network Design & Support – activities that focus on the design and support of local area networks, wide area networks, virtual private networks and the Internet ...	✓	✓	✓

Figure 37. Activities governance matrix for decentralized organizations (continued)

Activity (By Layer)	Enterprise	Business Units	Suppliers, Partners & Customers
Support Layer **Security & Privacy** – activities that focus on security architectures, authentication, authorization, administration and business resumption planning ...	✓	✓	✓
Procurement & Asset Management – knowledge that focuses on how to procure and manage technology assets ...	✓	✓	✓
Business Technology Acquisition Strategy – activities that focus on all aspects of the technology procurement and support process, including especially in-sourcing, co-sourcing and outsourcing ...	✓	✓	✓
RFP & SLA Development – activities that focus on the development of optimal requests for proposals (RFPs) and service level agreements (SLAs) ...	✓	✓	✓

Figure 38. Governance matrix for policies, procedures and regulations

Policies, Procedures & Regulations	Enterprise	Business Units	Suppliers, Partners & Customers
Healthcare Insurance Portability and Accountability Act (HIPAA) – which requires careful handling of patient records	✓	✓	✓
USA Patriot Act – which requires banks to use "reasonable procedures" to make sure that terrorists do not gain access to the financial system	✓	✓	✓
Sarbanes-Oxley Act – which requires officers of public companies to provide additional data to shareholders	✓	✓	✓
Basel II Accord – which rewards banks and other financial institutions for slid risk management methods and tools	✓	✓	✓
Fair & Accurate Credit Transactions Act – which requires uniform standards for credit	✓	✓	✓

Figure 38. Governance matrix for policies, procedures and regulations (continued)

Policies, Procedures & Regulations	Enterprise	Business Units	Suppliers, Partners & Customers
Security Breach Disclosures – which requires companies with customers from California to report every single breach of personal data	✓	✓	✓
Rules around data retention which include procedures around how long data – such as email – is retained and when it is destroyed ...	✓	✓	✓
Rules and regulations around the privacy of customer, supplier, partner and employee data ...	✓	✓	✓
Procedures around the violation of software licensing agreements ...	✓	✓	✓
Rules around intellectual property ...	✓	✓	✓
Specific procedures and emerging rules around digital rights ...	✓	✓	✓

As you inspect the matrices, you'll notice that governance resides squarely in hands of the enterprise in centralized organizations. This control occurs at all of the activity levels and all of the process and procedure levels – and especially the ones that require compliance. At the same time, all of the activities share governance among the centralized enterprise and its suppliers, partners and customers. This sharing is necessary and smart. As business becomes more collaborative and collaboration becomes more global, the need for cooperation among all of the participants in transactions is obvious. Make sure that your approach to governance acknowledges the importance of these "extensible transactions."

Figure 37 presents the governance guidance for **decentralized** organizations. While the recommendation is still to share some governance with partners, suppliers and customers, there's a bona fide sharing between the enterprise and the business units responsible for the many activities that together define business success. But business technology best practices suggest that there's a limit to the autonomy that business units should expect.

Figure 38 looks at regulatory and compliance governance.

How to Govern

It makes sense to have formal committees, steering groups and the like to make governance real. But what should be governed? What are the business and technology trends that require decisions about responsibility and authority (governance)? What's the government likely to do over time? Is more government regulation in the cards? Or, if the economy takes off again, are we

Figure 39. Governance planning matrix

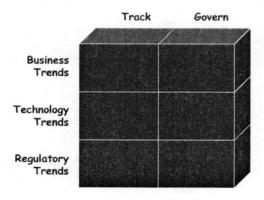

likely to suspend some of the reporting requirements that resulted from the excesses of the 1990s? It's important to launch a "watch" strategy, a series of activities that track trends in business, technology and government regulations. Figure 39 presents a real simple matrix that you might consider populating with activities, technologies and regulations. When completed, this matrix will help define your governance requirements.

If we fail to plan for governance we'll fail to govern. Effective governance is the result of environmental understanding and corporate efficiency. Figure 39 describes the range of the governance tracking challenge. It also identifies at least three watch areas: business, technology and regulatory trends.

There's an argument for defining governance broadly. It assumes that the world is changing, that governance and shared governance must extend beyond corporate firewalls because emerging collaborative business models require cooperation – which is, to come full circle – facilitated by shared governance.

Here are ten things to think about:

1. Governance does not stop in your parking lot. There are lots of internal and external governance issues driven by emerging collaborative business models that are redefining the scope of governance.
2. There are "layers" of business technology – strategy, applications, architecture, infrastructure and support – that generate activities that must be governed.
3. There are policies, procedures and regulations that drive your governance agenda. You have government produces at an alarming rate.
4. There are organizational structures – especially centralized and decentralized structures – that define optimal governance assignments.
5. Activities, policies, procedures, regulations, organizational structures – and therefore governance – are influenced by corporate cultures.
6. The assignment of governance responsibility should itself be governed by the development of matrices that list activities, differentiate among organizational structures, and identify the major policies, procedures and regulations that should be governed.
7. Business, technology and regulatory trends should be tracked to keep abreast of the governance requirements that must be satisfied.
8. Companies should convert governance requirements into realistic matches among their requirements, structures and cultures.

9. You should develop a governance plan comprised of activities, policies, procedures, regulations and structures that together constitute a comprehensive approach to internal and external governance and shared governance.
10. You need to "sell" governance to the powers that be.

Business Technology Management Convergence

Management is largely common sense. Do we agree? Come on. We get ourselves in trouble because we don't have defined processes or the discipline to enforce them. And some of us actually like ambiguity: it gives us power. Business technology management convergence is all about defined processes and discipline, whether it's the discipline to measure, develop business cases or actually do project management, and the discipline to standardize, manage funding or make the tough decisions about outsourcing.

All of this is about the convergence of collaboration and technology, the interrelationship between two high-velocity forces. What's always amazed me about business technology management is our ability to **know** the "right" thing, and our inability to **do** the "right" thing when the time comes to make the right call. Why is this? Why do management gurus continue to make tons of money telling us to, among other things, develop solid business cases, conduct serious TCO and ROI analyses, measure our environments and standardize? They keep coming up with new ways to motivate us, and new labels for old processes. Maybe good management is more about knowing ourselves and the people around us than the content of the decisions we have to make.

Anyone Want To Talk About All This?

The CEO ...

"Well, I guess a lot of this falls on my doorstep ..."

The General Counsel ...

"Yes, it does, but also at ours ..."

The CFO ...

*"Don't look at me: I've been telling everyone to get some discipline for years. No one wanted to listen – hell, you still don't ... I mean it's nice that we're looking at this stuff – **again** – but what's really going to change? You can still save money if you have discipline ... hell, we knew that three decades ago ..."*

The CEO ...

"We knew it before then ..."

The CIO ...

"Amazing ... absolutely amazing ... you people want great technology for as little money as possible ... you want total flexibility, scalability and security ... you want us to accommodate your every wish ... but because you're basically disinterested in technology, you spend very little time on technology management – except to yell at us when stuff breaks ..."

The CMO ...

"This is a boring discussion ..."

The CLO ...

"We should develop a training program around business technology management ..."

The CEO ...

"Why? We've been there and done that – about ten times ... that's not what we need ... we need to get serious about all this, really serious ..."

The General Counsel ...

"Do we have the right stuff?"

The CSO ...

"It's not our culture ... we're just not disciplined ... we don't even spend enough on security to satisfy our auditors ... why would we expect to be able to behave like real professionals?"

The Chairman of the Board ...

"The real question is about our capacity to change ... our capacity to step back and honestly assess ourselves ... and then take a deep breath and change the company ... can we do this?"

The CEO ...

"In other words, the last thing we need is another off-site meeting to launch another major corporate initiative ..."

The President ...

"Right ... we need something very different ..."

The CEO ...

"Where's this conversation really going?"

The Chairman of the Board ...

"Where do you think?"

Endnotes

1. CIO Insight Research Study, "ROI Overview," *CIO Magazine,* March 2002.

2. See Rachel Berg, "The IT ROI Roadmap," *Customer Support Management,* November/December 2001; Tracy Mayor, "A Buyer's Guide to IT Value Methodologies," *CIO Magazine,* July 15, 2002; and Greg MacSweeney, "Taking the Guesswork Out of Calculating Technology ROI," *Insurance & Technology*, November 2001.

Chapter VII

The Tough Conversation - It's Still (And Always) About People

What do we have here? We're going to the other side, to where civilized men and women seldom venture, to where the people who live and work among us are exposed. Here's the agenda:

- Who are you best people – defined in terms of intelligence, motivation and energy?
- What are the areas – like obviously collaboration and technology integration – that you want them to know and learn about?
- How do you keep nasty, stupid, arrogant, obnoxious people out of your company?
- How do you nurture and keep the best of the best, the people with knowledge, intelligence, experience, personality and character?
- How do you keep the culture clean?
- How do you groom 21st century business technology leaders?

Who Are These People?

We've talked about new business models, technologies, organizational structures and business technology management. Pretty good discussions. I don't think we left much out, though we probably didn't make too many friends among short-sighted product or service vendors or the good 'ol boys who really don't know all that much or work all that hard.

Now it's time to talk about human capital. Or stated just a little differently, let's talk about all the god damned problems you have with people.

And let's be very honest with each other. Not too many of us are good at this (though many of us think we are). We're also deeply suspicious about our internal hired help (a.k.a., "human resources"). And so it goes. If everyone wasn't so litigious, I could tell lots of stories about people who thought they were good at identifying talent but were incalculably horrible at judging either talent or character. Even more bizarre are the self-characterizations of these people as "expert" talent scouts. (Of course, the talent scouts also think that they're good managers, with extraordinary, if not magical, people skills – of course, the truth is just the opposite.)

We need people, and in spite of efforts to replace them with machines, the need keeps growing, simply because machines aren't creative and can't fix the problems they create. But we don't manage people well. We're inconsistent, subjective, biased and vulnerable to the same forces of nature that have been with humans since the beginning. (Yes, those forces.)

How much do you know about the people who work with you? How much do you know about what they do well, poorly and not at all? How do you motivate them? What do you expect them to learn? Is your company a "meritocracy" or do people win by any and other means? I remember visiting a data center some years ago and was struck by the number of people playing cards at 2:00 in the afternoon. When I asked what the hell that was all about, I was told that it's been going on for years, that cards were a kind of therapy for data base administrators and legacy applications people who – I was reminded – lead very stressful lives. OK, I thought, playing cards during breaks is probably a good thing. But they weren't playing during breaks. They were playing most of the day. Is this kind of thing all that unusual? Not as unusual as you might think! There are entitlement cultures in lots of industries and companies. How big a problem do you have? If you're a large organization, you might not know – or want to know.

How much progress have you left on the table because of professional or personal "friendships"? Do you play golf with a bunch of people who you've not fought with for a long, long time? Are you comfortable with people because they allow you to behave in certain ways? Like a jerk, for example? Would you want to be in a foxhole with these people? One measure of camaraderie is what happens after a meeting is over and a few stragglers talk about what just happened. If the chatter is negative, there's a disconnect brewing between those who lead meetings and those who attend them. If the negativity is entertaining enough, the number of stragglers will grow over time to the point where people attend the meeting just for the after-meeting chatter, the entertainment. Bad sign. There's also the "Scotty, beam me up" syndrome. How many times have senior people in your company said unbelievably bizarre things only to have the audience nod approvingly? Very bad sign.

The really hard part about all of this is us. We like or dislike, respect or disrespect lots of people – hell, everyone – based on a suite of biases that define who we are (and who we wannabe). Fortunately and unfortunately, the collection of these biases defines a corporate culture. Sometimes the biases are healthy, but often they're not. The problems on Wall Street, in board rooms and among senior management teams often stem from personal and organizational biases. How many times have you deliberately hired a senior executive who you know looks at the world completely differently than you do? How many times have you asked your team to think-outside-the-box and really meant it? Have you institutionalized a "devil's advocate" role in your company? People seek comfort zones no matter how much money and power they have.

Here's what I think. You tell me if it's profound (or not).

If we like ourselves – even if only superficially – we surround ourselves with people like us. If our "team" is too diverse we get nervous. We work with certain people because we have to, not because we have open minds about the potential quality of their contributions. If we're male, we resent female and other explicit or implicit quotas. If we're female, we're always looking for the glass ceiling. There's a weak correlation between the people we **like** most (note that I didn't say **respect** most) in our companies and their professional contributions to the company's health and welfare, and if you don't believe this then you're anything but objective. There are lots of people better qualified for key positions in your company: you just don't know any of them – or if you do, you don't like them as much as you like the comfortable incumbents. If you've been working with a large number of people "for years" your organization is under-achieving.

The number of people in your company that are really smart about your vertical business processes is very small, so small in fact that you single them out as "high potentials" (while implicitly informing all the others that they are "low" – or have "no" – potential). The number of people in your company that are really smart about computing and communications technology is also very small – probably smaller than the number of smart business professionals you have. Why? Because the pace of technology change has been faster than the pace of business change. The number of people that are smart about your vertical industry and technology – and in your company – is miniscule.

The buddy system – an old constraint – has been around since people began to congregate. It's naïve to assume that we could break such an embedded pattern. We're taught selectivity in sports, fraternities, sororities, professional associations, political parties and even our churches. We learn all this early and often. Aspects of bias are useful, such as the loyalty and teamwork that it generates. But when it comes to corporate performance we need to make more objective assessments of what our colleagues can and cannot do – and, as we'll discuss in a moment – whether or not they're jerks. Business technology convergence is complicated and dynamic. We have to be careful with the people decisions we make around collaboration and integration.

A little reality: all of us know that even if we had more brilliant well-adjusted professionals than idiots and jerks in our companies, we'd still be stuck with a fairly large percentage of our workforce. How come? Well, it's the way of the world. It's really hard to fire people with de facto "tenure" or people we've known for lots of years. Why do you think boards like to fire under-performing CEOs? The new management team has no loyalty to the old regime. It can pursue whatever variation of scorched earth restructuring it wants. But if there's no wholesale change at the top, we have to make due with the middle and bottom we have. This means that serious attention should be paid to how we get the most of the little we have to work with. After you figure out what you've got back at the ranch, you can work around the constraints.

So where does all this leave you? Am I accusing everyone of bias and vested self-interest? Do I believe that no one's capable of hiring and nurturing smart people? Of course not, at least not completely. I would argue, however, that it's actually getting easier to develop collaborative strategies but much, much tougher to execute them. **The old constraints are still with us but execution complexity has increased considerably.** It all starts with the amount and quality of human capital at your disposal.

How's The Team — Really?

Here's something to get you started. Take a look at the matrix in Figure 34 and then locate your people. How bad is it?

Let's have some fun with the cube. If you have lots of smart people with no energy or ambition, you have a problem (unless you're in a vertical industry that rewards inactivity). But it's actually more dangerous if you have dumb people with tons of energy and ambition. The military has known for centuries that officers can be smart **and** arrogant, but never arrogant **and stupid**.

Clearly, the goal is a bunch of smart, energetic and appropriately ambitious professionals each with a set of known strengths and weaknesses. **Known** is the key word here. If you know what people do well and poorly then you can assemble successful teams. If you don't – or the individuals don't see their strengths and weaknesses – you'll spend lots of time fixing dysfunctional teams.

But what does **smart** mean?

Figure 35 presents at least three kinds of knowledge which, of course, need to be integrated. Generic, structured knowledge includes facts, concepts, principles and formulae that describe what things are and how they work. Finance

Figure 34. The people placer

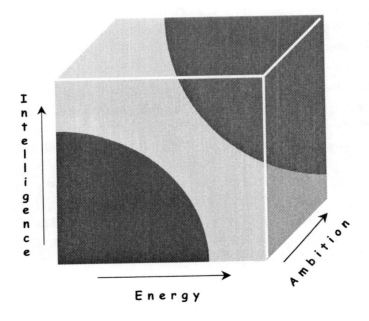

Figure 35. Integrated knowledge

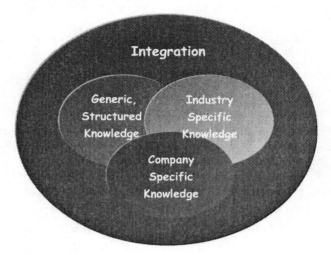

is a good generic, structured field. Computer Science is another one. College students major in these fields. Industry-specific knowledge comes from different sources. A little comes from colleges and universities, but most of it comes from on-the-job experience, training and industry certifications. Company specific knowledge comes from time spent in the trenches of your particular corporate domain. We often place great value in institutional memory, but be careful about how easily distorted such memory becomes. People with political agendas are terrific at re-writing history to match their current vested interests. (In large organizations, company-specific knowledge often includes maps that describe where the bodies are buried.)

And what about the other members of your team? Suppliers? Partners? Even customers? What do they know? Lots of our efforts to collaborate and integrate will depend upon the intelligence, energy and commitment of our employees, suppliers and vendors. What if they're "low potential"? Seriously, how good is the whole team? Just as you need to assess your employees you also need to assess the strengths and weaknesses of your collaborative network. If you find some serious problems you need to fix them. Sometimes you can fix problems with a new collaborative process or a piece of integrated technology. Sometimes the problems are more fundamental and specific people need to be replaced.

When we talk about "smart," we're talking about depth in the three knowledge classes as well as the ability to integrate them into insights, inferences and

Figure 36. Knowledge → intelligence, energy and ambition

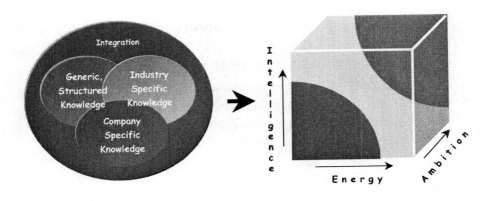

decisions – all suggested by Figure 36. But "intelligence" is fed by integrated knowledge combined with raw intellectual horsepower. Energy and ambition are measured independently. Have you ever tried to measure them?

OK, time to talk candidly about people. Some of your people are really smart and some are not. Some work at understanding existing and emerging collaborative business and technology integration trends, and some don't. Some even work at increasing their natural energy levels, but most don't. And some work to compensate for their weaknesses by leading with their strengths.

Some want your job. Some are clueless. Some are evil. Some are sweet. Some got where they are mysteriously; some really earned it. Who are the "keepers"?

If you're in charge of this zoo you need a large bag of tricks – and the will to frequently reach into it for just the right one. There's not too much you can do about raw horsepower: we're born with the basics. But there's a lot you can do about the availability and insertion of knowledge, especially industry – and firm-specific knowledge. Given what we've already discussed, here's a short list of knowledge that would be good to command:

Generic Structured Knowledge

- Solid basic knowledge about business, management, finance, accounting, marketing, technology, and all of the areas that enable the acquisition and application of more specific (industry and company) knowledge.

Industry Specific Knowledge

- **Collaboration** requires knowledge about the inter-connected marketplace and inter-networked companies, knowledge about what happens inside and outside of companies, and the collaborative mindset.

- **Customization and Personalization** demands knowledge about mass personalization, behavioral models to correlate online and off-line behaviors, wireless personalization and personal and professional CRM, among other related areas.

- **Supply Chain Management** means knowledge that includes supply chain concepts, models and tools. Integrated supply chain management (by vertical industry) would be a central focus here along with the technologies that enable supply chain management as well as SCM standards, technologies (such as exchanges), and some of the leading SCM platforms.

- **Automation** calls for knowledge about intelligent systems technology and the application of that technology to personal and professional automated transaction processing, monitoring, e-billing, and the like, including methods (neural nets, fuzzy logic, expert systems) and how these methodologies can be embedded in tools and applications.

- **Optimization** demands knowledge that looks at major technology and business processes and how they can be optimized with a variety of models, tools and technologies as well as the need for integration, interoperability and synchronization and how optimization becomes the nexus for productivity and profitability. Optimization concepts, models, tools and technologies can be applied to technology performance (network optimization, for example) and business process performance (applications design and development, training, customer acquisition, etc.).

- **Applications Architectures** requires knowledge that looks at how mainframe (single tier), client-server (2 tier) and Internet/Intranet (3 tier →n tier) applications have changed and what the trade-offs among the architectures (defined around flexibility, scalability, reliability, etc.) are.

- **Messaging and Workflow** includes knowledge that examines the platforms that support all varieties of communication and how communications technology enables communication and transactions among employees, customers and suppliers inside and outside of the corporate firewall.

- **Database Management and Analysis** calls for knowledge that positions data, information, knowledge and content – of all varieties (static, dynamic, text, video, etc.) – and how it can be managed for alternative purposes, as well as data, knowledge and content management platforms, next generation database management applications (especially object oriented DBMSs).

- **Integration and Interoperability** calls for knowledge that describes the technical requirements for making disparate, incompatible applications, standards, data, platforms and architectures communicate with one another, while focusing on enterprise applications integration (EAI) and Internet applications integration (IAI), wrapper/glue technologies like XML, as well as more conventional middleware. The knowledge should focus on the need for – and objectives of – integration and interoperability including cross-selling, up-selling, customer service, alliance building, etc.

- **Business Technology Metrics** depends upon knowledge designed to introduce professionals to ROI, EVA, TCO (and other) models for assessing business technology effectiveness. Business case development and due diligence should also be included here.

- **Security and Privacy** calls for knowledge that examines the concepts, models, tools and technologies that enable security architectures, authentication, authorization, administration and business resumption planning. The technologies would include encryption, biometrics, PKI and smart cards, among others.

- **Business Analytics** cannot be done without knowledge about the processes and technologies that yield insight from sales, marketing, customer service, finance, accounting, technology infrastructure and competitor data, as well as knowledge about the forms that such analyses can take.

- **Project and Program Management** requires knowledge about project management processes, methods and tools as well as program management processes, methods and tools. The range of areas would include several varieties of business technology project management and several varieties of program management including business technology acquisition strategies, managing outsourcing, service level agreements, etc.

- **Partner Management** calls for knowledge that includes approaches, methods and tools for managing relationships with distributors, re-sellers, service providers, etc.

- **Regulatory Trends** cannot be analyzed without knowledge about regulations and regulatory trends in specific industries and hit lists for tracking legislation that could significantly impact business policies, processes and procedures.
- **Business Technology Acquisition Strategies** call for knowledge that examines all aspects of the technology procurement and support process, including especially in-sourcing, co-sourcing and outsourcing.
- **Professional Communications** includes knowledge that helps people understand the form and content of professional written and verbal communications especially as it involves the communication of business technology.
- **Manufacturing** depends upon knowledge about your industry's manufacturing processes and technologies.
- **Distribution** depends upon knowledge about the industry's distribution practices.
- **Service** entails knowledge about the industry's approach to customer service.
- **Business ←→ Technology Convergence Strategy** requires **industry-specific** knowledge that examines methods for developing and assessing business technology strategies in specific and converging vertical industries. Some of the models and methods that would be included are scenario planning, decision modeling and alternative futures development. These methods would then be linked to major technology investment decisions around applications, communications, data, etc. Such knowledge would help professionals understand the relationship between vertical collaborative business strategies and integrated computing and communications technology.

Company-Specific Knowledge

- **Business ←→ Technology Convergence Strategy** calls for **company-specific** knowledge that examines methods for developing and assessing collaborative/integrated business technology strategies in your company, including knowledge about the current scenario that's driving your collaborative business strategy and the plan for integrating computing and communications technology in your company. This knowledge is your

corporate mantra, your *raison d'etre*: all of your employees should drink from this fire hose.
- **Alternative Business Technology Scenarios** require knowledge about competing scenarios to your primary one, as well as the scenarios for your conventional and unconventional competitors.
- **Mission and Values** demands knowledge about your company's core mission and the values to which it subscribes.
- **Competitor Intelligence** depends upon knowledge about specific competitors including information about their sales, marketing, profitability, strategy, etc.
- **Policies, Procedures and Discipline** needs knowledge about the way your company operates, how it's organized, how it makes decisions, how it rewards and punishes people, how it deals with supplier, customers and partners, what it likes and dislikes, and insights into its "personality." Note that this is very tricky knowledge to create and communicate. At the same time, there's no more important knowledge for employees to have.
- **Relationships** depend upon knowledge about the relationships the company has with its suppliers, customers, partners, benefit providers, etc.
- **Culture** requires knowledge about how your mission, values, policies, procedures and discipline together comprise your culture including "stories" of the company's greatest successes and failures.

How much do your people know about this stuff? When we talk about education and training, this is the range of knowledge you want your people to have.

Now let's talk about the "jerk factor."

If you're new to an organization – as I've been several times in my career – after a week or so of "observation" you begin to make mental lists. One of them is a list of the people who are so far over the top that you find yourself slipping into a state of buyer's remorse, wondering how you could have been so stupid to accept the new position. (True story: one of the organizations I joined - to which I was a consultant – offered me a terrific position, which I accepted. After a week or so on the job the people with which I consulted stopped me to note, "We thought you were smart, but clearly you're not ... if you were, you would never have accepted our offer.") People of course fall into all sorts of categories. Some are hopelessly rude and arrogant. What do we do about these people (that have buddies just like them all over the place)? What do we do about people that disrupt and undermine? People that complain all of the time? People that have nothing to offer but bitterness, anger and jealousy?

People can be smart, ambitious and energetic, but also arrogant and caustic. Who do you want on your staff? To whom do you entrust major business technology initiatives? How do you make the trade-offs?

When companies are making lots of money, people (and companies) suffer fools amazingly well, but when times get tough tempers and patience grow short. When times are good you should find what the jerks do best and isolate them accordingly. When times get tough you should prune them from your organization. Ask your high performers who they avoid and why they avoid these people. A consensus of opinion usually represents reality. Go with it.

The final word is ethics. I have always told my clients and students that ethical behavior is not always an optimal short-term course of action, but it's always a long-term optimization strategy. In short, good ethics is good business, especially if you intend to stay in business. We've already established that customer relationship management is a process before it's a technology. It's

Figure 37. Smart productive jerks?

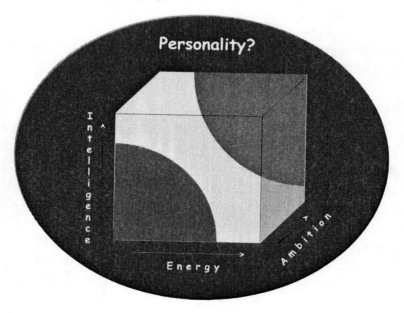

hard to imagine that companies would not aspire to ethical CRM – especially if they want to keep their customers! Do companies spin stories about their products and services? Of course, that's normal – and built into consumer perceptions of value. But there are ethical lines that should not be crossed. Part of your due diligence on especially senior people should focus on their character. Why so much attention on the senior people? Because the senior folks set the ethical tone for the company. If junior people see their leaders behaving unethically, many of them will follow suit, thinking that imitation is the sincerest form of flattery (and therefore a path to "success" in the company). Pretty soon – after the ethical employees have left the company – all that's left is unethical leadership supported by unethical managers and worker bees, and a bunch of other people too cowardly to either speak up or leave. Terrific.

Stay Sharp

Another thing to keep in mind: the pace of business technology change is generating requirements faster than employers and employees can satisfy them. This is stressing "learning organizations" to the point where they have to invest in serious education and training programs for their employees. "Corporate Universities" are springing up all over the place and the number of new content titles is growing by leaps and bounds. All of this effort, however, must be targeted at coherent strategies. If they're not, then all of the work to keep employees current may be misdirected: training requirements should be derived from collaborative business technology strategies and tactics – not the other way around.

Remember the discussions we had about "core competency"? Let's think about it in terms of business technology convergence. Should it be on the list? Is anybody there? Of course. Business technology convergence should be one of the top three competencies in your company.

This isn't the first time you've assessed your core and non-core business models and processes. You've assessed them to determine strategic technology investments and to determine how much outsourcing you should consider. This continuous core/non-core assessment should now yield some insight into what skill sets you need and how to acquire them through education and training.

If you haven't outsourced lots of business technology, you have some serious training (and education) requirements to satisfy. Even if you've outsourced some work, you'll still have to keep your business technology professionals current. If you have lots of business technologists, then you've probably institutionalized a continuous training process – and hopefully you're measuring its effectiveness.

But there might be problems with your strategy. First, let's make sure that we're talking about the same things. Your employees need to understand collaboration trends as well as the technology trends that will enable collaboration. They also need to understand the principles around which key business technology decisions should be made. **Principles**, you say? Yes. Because while you can train people to perform specific, even complicated, tasks, you cannot train them to think creatively with informed perspective. Or can you?

New hires get courses about your industry, your company and technology. If you still develop lots of applications (for whatever inexplicable reasons) you probably already have courses in systems analysis and software engineering (your core competency assessment will have identified such courses as core to your continuous learning technology requirements). It's important that learning be continuous and current – so long as the content is right.

The above list of knowledge areas should drive your education and training strategy. You should especially offer courses in industry- and company-specific knowledge areas. If you treat your knowledge investments in people as long-term, then you should offer educational benefits as well as the more obvious training ones. If you don't offer educational benefits, your employees will assume that you're thinking about them as short-term assets – and plan accordingly.

The delivery of content – and the infrastructure necessary to do so – must be specified. While face-to-face content delivery always makes sense, there are times when some flavors of distance education also make sense. Face-to-face

Figure 38: Knowledge delivery options

Place \ Time	Same Times	Different Times
Same Places	"Face-to-Face"	"Asynchronous"
Different Places	"Distributed-Synchronous"	"Distributed-Asynchronous"

delivery obviously requires an instructor, space and transportation. All of the other models require some delivery technology and a technology infrastructure to support continuous learning. (But remember all you really need is access to hosted content. You don't need to host and manage learning content yourself.) Take a look at Figure 38 – it presents your options.

What Makes Sense?

Face-to-face (FTF) delivery has advantages – and disadvantages. Some knowledge is best communicated FTF, while some can delivered over networks. The real decision is what to deliver synchronously and what to deliver asynchronously.

Some rules of thumb. When "answers" are judgmental, FTF is best. When cases need interpretation, when strategies are gray and when arguments enhance learning, it's better to have humans in the same room. But when the materials are unambiguous and predictable – more factual than interpretive – distributed asynchronous learning is just fine. Use your judgment: knowledge can be delivered in several ways and sometimes even redundantly.

In order to support continuous learning you'll have to create or buy content in the right knowledge areas. You can do this in-house or outsource it. You can make arrangements with training companies, universities and educational consortia, and connect instructors with students regardless of their location or

Figure 39. Collaborative learning network

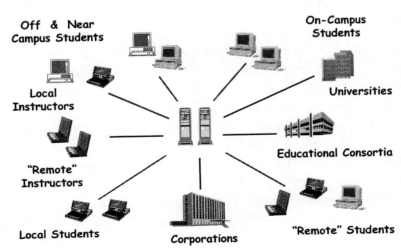

mobility patterns. There are lots of content providers who would be willing to work closely with you to develop custom knowledge. Asynchronous learning networks can be developed very quickly. Look at Figure 39 to see how these things can be configured.

The above knowledge list is a snapshot of what's relevant today. There's no question that the list will change. Ideally, you'll find a content provider with which you can work closely – and a reliable stable of online and FTF instructors.

They're Alive, They're Alive!

Who are you looking for? If you buy the collaboration/integration arguments were making here, then you need people who understand these trends – smart people who are housebroken. You can also evolve them.

Let's start with the search. Networking is the best way to find the right people, especially if your corporate culture is consistent and sane enough to sustain an image that translates into opportunity for bona fide business technology professionals or professionals with potential. (If your company y has a bad reputation you need to fix that problem before going out there with a happy face.)

I recommend a three-pronged strategy (I always wanted to say that). First, I'd directly connect with local, regional and national universities. Here's how I'd optimize these relationships:

- Volunteer to work with faculty and administrators on curriculum content: for every course with meaningful content you can reduce your knowledge requirements proportionately. Community colleges and four-year colleges and universities would love to hear from you. Really. OK, some faculty react badly to "hi ... we're from industry and we're here to help," but some others – the ones who really care about connecting their teaching and research to the real world will welcome you with open arms (especially since they'll also expect you have deep pockets for funding their applied research and curriculum reform). The way you manage this relationship is through contact that's proactive. Throw ideas out for their consideration. Develop new course ideas, new ways to teach applied content, and offer access to your companies for field trips, faculty sabbaticals and – especially – internships.

- Internship programs are the most efficient "try-before-you-buy" recruitment methods out there. Why wouldn't you have one? You get to handpick college and university juniors and seniors, pay them relatively little (and sometimes even nothing), and assess their abilities on the job, in your own house. What's not to like about this recruiting approach? Help local and regional universities develop internship programs targeted at placing people in companies like yours – especially after you've influenced the curriculum. This one's a no-brainer.

- Connect with faculty interested in applied work. Many professors are really smart about the new tools and techniques as well as trends in your vertical industries. It's their job to think about this stuff. Tap into their brains with consulting contracts (they're cheaper than the commercial people) and sabbatical opportunities. The same faculty can also tell you who the best students are.

Next, I'd develop a **human capital intelligence operation** designed to find and seduce your competitor's best people. I'd do this because their level of generic and industry specific knowledge is already high and chances are they'd integrate easily into your company. But what about professional ethics? Well, I'm not suggesting that you plant lipstick cameras all over your competitor's offices, but I am suggesting that you build a database of your competitors' best people and contact them from time to time about the possibility of moving to your company. This is a really efficient way to recruit, since most of these people can hit the ground running wherever they land.

Third, I'd recruit through my vendor relationships. I'd watch the people working on your projects and when you spot some really good ones, I'd approach them about moving to the other side of the outsourcing relationship. Turns out that lots of these people would like to move to stable environments, especially in situations where they're required to travel only a few times a year. This is not raiding: it's natural market evolution. Consulting companies in particular often like it when their semi-senior people leave, since they get to replace them with cheaper people (who generate higher margins). Vertical companies also like to bring consultants in to work directly for them. They often bring new ways of thinking about old problems and while many of the problems are persistent and to an extent unsolvable, they still like new perspectives – that they think they can get from career consultants.

Why no "conventional" recruiting? Because it's expensive and unlikely – compared with the other methods – to yield the right results. Look, you're shaking trees to see what falls out. I'm suggesting that you shake the right ones at the right time. Going to trade fairs and trolling in newspapers and professional journals and magazines are terrific ways to tell the world you're hiring people, but awful ways to find people with the right knowledge profiles. If you have people traveling around looking for people, pull them back to the ranch, reassign them to other recruiting methods, or reduce your overall salary expense by pushing them to the other side of the recruiting river.

In addition to the knowledge/intelligence/ambition/energy filters, I'd also apply the jerk and character filters each and every time I think I'm in love with a prospective employee. Everyone knows that personalities explain a ton of productivity, especially in outward facing businesses. Avoid rude, goofy, nasty people. They'll wreak havoc in your companies (as if you didn't already know this). A healthy amount of your candidate due diligence should always focus on personality.

How About Growth Hormones?

The problem with growing people for the long haul is that it's unlikely they'll be with you for the long haul. Why is this? Because over the past couple of decades, we've created the expectation that people will have multiple careers and will spend three to five years with individual companies before moving on. This complicates the allocation of human capital to human capital. Recruiting people with deep industry experience always makes sense, especially if it takes a couple of years to train novices up to speed. The ideal employee is someone who has generic, structured knowledge, industry-specific knowledge and experience and the ability to learn your company's policies and procedures quickly. Oh yeah, and they should be nice, too.

A word about looking way beyond the border for employees: desperation correlates with distance. If you find yourself looking to import talent from a couple of continents away, then you're looking too far from home, way too far. Why do I say this? Is it because there aren't smart people in India, Russia or Korea? Of course not! Some of the smartest people in the world are from these and other countries. So why not tap into these human capital markets? The answer is simple: overhead and hassle. It takes lots of time and effort to make it work right and when it does you still have to invest heavily in training about your vertical industry and company. In other words, you can import generic structured knowledge, but it's relatively hard to import people who can hit the ground running in your industry and your company. Obviously there are labor shortages in some generic knowledge areas, like computer science, and it can make sense to recruit people from wherever to fill these positions, because generic skills are transportable. But if your requirements are industry- and company-specific, then you should recruit closer to home.

Keepers

Whatever happened to **mentoring**? I don't mean the casual bonding that sometimes occurs between newer and older employees, but formal mentoring programs designed to accelerate the company-specific knowledge acquisition and communication process. When new people come into an organization they should be given a week to find a mentor to not only show them the proverbial ropes, but to help them plan their careers in the company. Mentors should be assigned to work with new hires for what equates to a day a week, for at least several months. After that, the mentoring should continue for as long as the employee and mentor are with the company, or until the relationship needs to be changed because of the trajectories of either career. A day a month for life sounds about right.

Mentoring is like continuous orientation. It increases the probability of a successful employee/company match and therefore helps with retention – and succession. The worst thing that can happen is to spend lots of time and money recruiting people that leave before they make any useful contributions to your company because they were lost somewhere in the shuffle. Formal mentoring is good business, and in order to make it work, positive and negative incentives should be used to make the programs viable.

What **incentives** do you use to keep good people? And how do you use them to send the right messages to the people you don't want to keep? There are lots of arguments here. Some think that the quintessential incentive is money, that no matter what else you offer there better be enough cash (in various forms) to please your star performers. There's a lot of wisdom here. People need to buy food, educate their kids and pay off their homes. One thing's for sure. If you underpay your top performers, you will lose them. We can argue forever about how much is enough, but if you don't pay it you'll lose people (to the competitor that has a database of your good people). So you have to find the right number and stay just above it for as long as you want to keep them on the job.

But money's not the only incentive. Increasingly, evidence suggests that environments that respect their employees and offer them the right learning opportunities keep their employees. Trust results from a mutually respectful and beneficial relationship between employer and employee. Profound, huh? Actually, while we all pay lip service to platitudes like this, they do keep us balanced – especially when meritocracies lose to golf handicaps.[1] What does this mean? If there's one aspect of a corporate culture that demoralizes employees at all levels, it's the perception (which is too frequently reality) that factors other than merit determine rewards. You've seen it and I've seen it. Frat boys, sorority sisters and golfing buddies who are anything but brilliant get promoted and rich because of who they know, not what they know or how they perform. When this kind of reward structure exists, it infects organizations at all levels. People become cynical, angry and disenfranchised when they believe that no matter how hard they work, how right they are or how well they perform, they won't be appropriately rewarded. So what happens when golf handicaps drive wealth creation? Several things. First, given the message that's sent loud and clear to the troops, the get-along-go-along culture will reduce your overall competency to mediocrity. Many of your employees, in other words, will adapt to the rules of the game that the buddy system plays by. They won't rock the boat, think outside the box, or – God forbid – challenge authority, because they understand that if they piss off the ruling boys – or gals – they'll never get rich. So they begin to spend more time working on their relationships with the ruling elite than with customers, suppliers or partners. The obvious result here is that business suffers. Next, the star performers who really want to improve the business – and who are uncomfortable with good 'ol boy/gal rules – leave your company to work for one of your competitors (who may or may not play by the same rules). Third, the company will eventually collapse under the weight of these rules if they continue to grow in number and complexity or if they spin out of control into what we've recently seen in the form of corporate anarchy, arrogance and irresponsibility.

Look, you know what I'm talking about here. If you stay closer to meritocracy than other reward structures, you'll create a culture that people want to join and where good people want to stay.

So what are the incentives that work? Here's a list of winners:

- Published compensation scales tied to specific performance metrics.
- Highly competitive benefits including flexible retirement accounts.
- Opportunities to assume more responsibility with authority.
- Opportunities to learn new things.
- Rewards based on merit.
- Mentoring.
- Sabbaticals.

Sabbaticals? You bet. When key people work really hard for a long time with consistently impressive results, you need to occasionally give them a rest. Is this that complicated? Look at the companies that offer sabbaticals and see what kind of loyalty they generate.

Ambiguity is bad. Specificity is good. If you're trying to build a meritocracy you need to write down what it takes to succeed and stick it on your Intranet for everyone to see. Employees need to know precisely what the organization expects them to do and how well it expects them to perform, and what it will give them when they perform well. This requires some planning, communication and the dreaded performance review.

Performance reviews have been around for a long, long time. They are some of the most political processes in your company. Some of your employees are so good at gaming reviews that actual performance has little to do with an employee's assessment. In highly political organizations, people spend a lot of time figuring out how to game performance reviews. In meritocracies, people spend time organizing and presenting performance evidence.

How should you do this? First, publish the process and the outcomes, which range from promotions, raises, bonuses, new responsibility, demotions and dismissals. Each year employees should participate in the development of performance objectives which should be used at the end of the year to assess how well the objectives were met. The employee's immediate supervisor along with a two person "independent" board should be involved in the review. Am I crazy? Am I suggesting that we get people to agree on what they should accomplish during the year and then review their progress? That three people can do this "objectively"? If the culture supports all this, the answer is yes. But if it doesn't, forget about it.

Business Technology People Convergence

It's always about people. Some people – perhaps you – think that all you have to do is get the "people part" right and the rest will take care of itself. There's a lot of wisdom there. Certainly the converse is true – several nasty idiots – especially in high places – can ruin a company. Unfortunately, it usually takes a long time, during which a perhaps once healthy culture is tortured into submission by corporate sadists.

The business technology convergence people piece is especially complicated because of the very different personal and professional places business and technology professionals come from: because of their disciplines and experience, they often see the world quite differently. One field's best practices are another's goofy elixirs. What to do? As we discussed in Chapter V, the first step is to tear down the walls around professional practices, and integrate – and then converge – as many of them as possible (which is why, for example, I recommend killing off the "CIO" and "CTO" titles in favor of enabling responsibilities that converge with collaboration and support responsibilities).

As with business technology management, here too the essential requirement is discipline. While it's important to know about different kinds of knowledge and to track learning requirements in your industry, it's much more important to erect high gates around your company – gates designed to keep nasty, stupid, arrogant, obnoxious people out. If your screening discipline breaks down, you'll inject yourself with infectious agents that will eventually kill your culture and perhaps even the company.

Knowledge, intelligence, experience, personality and character: five things I need to know about you – and everyone else in the company. These are inviolate. The scary exercise is to reverse engineer your hiring over the past couple of years with reference to the five filters.

Ten years ago business technology leadership could be defined around desktop operating system upgrades. Today that same project would probably require leadership around the development of the RFP and SLA necessary to outsource the activity. As supply chains compress, leaders are necessary to optimize the business technology partnership.

Impact and influence are changing as well. In the 1990s it was more than enough to report on a successful implementation of a back-office accounting system. Today the only meaningful performance metrics are business value metrics. Did the project increase our market share? Our profits? Are our customers happier now that we invested in CRM?

Communications are changing. Business cases must now be in the language of business, not techno-speak.

Does all of this mean that operational technology is unimportant? Of course not. But business technology leadership is about above-the-line excellence, not below-the-line efficiencies – which are expected, just as compliance to Sarbanes-Oxley is expected.

Another way of thinking about 21st century leadership challenges is to ask yourself – and your staff – if they think that the jobs they're performing today will persist into 2010. In a recent survey I took among over 100 technology professionals across three companies (one public, one private and one not-for-profit), **not one** director-level (or higher) technology professional believed that he or she would be doing the same thing they're doing today just five years from now. If this anecdotal data is correct, we'll all have new jobs by 2010. What will they be?

In the midst of all these changes, what should business technology "leaders" do? Among other things, they should:

- Build Collaborative Business Scenarios
- Track (Only) Technology that Matters
- Identify Business Pain and Pleasure
- Organize Adaptively
- Manage Infrastructure
- Communicate
- "Market"

They do these things because they all – in one way or another – connect them seamlessly and holistically with the business models and processes that determine competitive positioning, market share, growth and profitability.

Business technology leaders should focus on business models and processes before they focus on technology infrastructure or applications. Macro business trends should occupy their time, trends like e-business, customization, personalization, and supply chain planning and management – all suggested by Figure III. Leaders get their companies in the upper right hand corner of the cube.

Scenario planning identifies the drivers of change, their velocity and the confidence we have in their validity. Good scenarios also quantify the uncertainties that complicate planning.

But what good scenarios really do is profile marketplaces and profitable transactions. They also identify constraints. They are compasses that influence the direction that strategic decisions take.

Business technology leaders develop, package and "sell" business scenarios. They work with the business to profile "as is" and – especially – "to be" business models and processes. They become champions of the scenario development process. They brand themselves as business-models-first-and-technology-heroes-second.

Leaders also track technology trends – especially trends that matter to the business. Trends that matter include all technologies that can impact business, not just technology "concepts" or even "prototypes." Examples? The semantic Web – the intelligent Internet – is a tremendously interesting concept – but it's a long way from implementation. Real-time synchronization and real-time computing, generally, are also fascinating concepts but, again, we're some years from away from productive implementation.

Utility computing, on the other hand – the technology acquisition and support model that uses the electricity model to describe its pay-by-the-drink approach to technology acquisition – is emerging as a prototype with some potential – though it's way too early to commit to a major investment in whole technology subscription models. Similarly, grid computing is showing some promise, as is Web Services technology, thin clients and the newest voice recognition technologies. These all bear watching so long as they map onto the business scenarios that the same business technology leaders develop.

Leaders also track dominant technology standards. Would you have invested heavily in Bluetooth wireless communications technology two years ago? Would you have done the same in Web Services in 2001? Are you tracking service-oriented architectures, or have you already moved to event driven architectures? Who's watching RFID standards? Leaders do a couple of things here. First, they watch the standards power brokers. Can any of us deny the impact that IBM's decision to support Linux had on the adoption of the operating system? Wal-Mart will yield tremendous power over RFID standards (as it has in the collaborative forecasting and replenishment, inventory and supply chain management areas). As the technology industry continues to consolidate, the number of companies with standards-setting power is actually shrinking, which is good news for leaders searching for direction. The second thing that leaders do is map the trajectory of standards onto their business scenarios. How will Wal-Mart-influenced RFID standards affect your business? How will IBM's support of Linux change the way your company thinks about server OSs?

Business technology leaders should see the world through the eyes of the business. They should speak the language of business. But most importantly, they should focus on the pain that business managers feel. The really good ones keep a running list of the most difficult problems – the sharpest pain.

Business pain comes in many forms. Some comes in the form of cost control, such as headcount and overhead cost reduction. Other pain relief comes in the form of improved business response and control, such as improved management effectiveness, employee productivity and supplier relations.

The search for business pleasure should also occupy a leader's time and energy. Some pleasure includes revenue generation, up-selling, cross-selling, organic growth, acquisitive growth and – of course – increased profit.

The whole pleasure/pain exercise focuses on business success. It also focuses on what individual business professionals will personally find exciting – and rewarding. Leaders understand what makes people heroes, what the organization values.

Figure 40 identifies three paths in the alignment-to-partnership journey. We have to appreciate business pain and pleasure, we have to become more than just credible, and we have to define business value around strategy. If you

Figure 40. Paths to business technology partnership

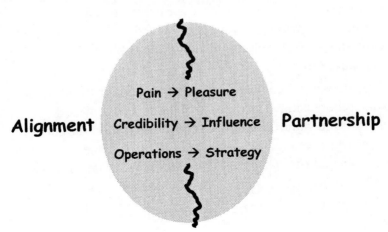

understand these paths, you can define the business technology relationship – redefine it, in fact.

Let's assume that you understand where the business feels pain, and how it would define pleasure. Remember that the business expects technology to reduce its pain – almost always defined around cost reduction. But it's more than that. Business managers also worry about their supply chains, their competitors, their manufacturing, distribution and, of course, their margins. The technology agenda needs to speak directly to their pain points – which, when relieved – can become sources of wide and deep pleasure. If you become a dispenser of pleasure as you reduce pain, your credibility will rise – which will reveal the second path to business technology partnership.

Hopefully, when technologists walk into a room the business managers don't run for cover or – worse – attack them mercilessly for their sins (network crashes, Web site debacles … you know the drill). Nirvana here is influence – defined in terms of how the business thinks about how and where technology can help. Does the business respect you enough to confide in you, to commiserate with you, to invite you to brainstorm about its strategy? Who do you drink beer with?

If you're influential, you can shape both operations and strategy. If you get operations straightened out, you can spend most of your time – with your new partners – thinking about competitive advantages, revenues and profitability. There's no better place to work, no better way to spend your time. Leaders seek this partnership, this influence.

So if you achieve the partnership, what do you give up? A little bit of yourself, a little bit of your technology experience and a little bit of your credibility with your legacy technology peers. What's that? Yes, because true partnership means that some people get a little less of your time and interest than they used to, that you should probably no longer play cards with the data center crowd, and that you'll have to start reading all new trade publications. You'll also have to buy some new suits.

Lots of companies are decentralized these days, though the number that are re-centralizing is increasing. The essence of the centralization/de-centralization dance spins around value of shared services. But it's also about discipline and governance. Many companies have had a difficult time standardizing their infrastructures and processes, so difficult in fact that they've resorted to extreme outsourcing.

The key going forward is to define the business technology organization as though there's one – not two – organizations. Operational infrastructure should be managed transparently, almost as though it doesn't exist. While this is not to say that it's unimportant (see below for just the opposite argument), it is to affirm its relative unimportance compared to strategic, customer/supplier/employee-facing projects, programs and impact.

The governance around all this should be clear, consistent and unambiguous. This will be the toughest battle you fight, but once the governance is set it becomes easier and easier to get things done. If you handle governance poorly you'll find it nearly impossible to organize business technology effectively.

Organizational leadership focuses first on the governance of business technology resources, investments, responsibilities, principles and priorities. The business value of technology should be the primary governance philosophy. Below-the-line infrastructure and support should be shared across the organization regardless if the business structure is centralized or de-centralized. Strategic applications should be identified by the business regardless if the structure is centralized or decentralized. Enterprise architecture should be jointly owned by the lines of business and the infrastructure support provider.

Reporting relationships are always complicated, especially in decentralized organizations. The key is to organize around the business technology layers that share decision-making authority through explicit governance (and business technology councils for exceptions handling). Flexibility is essential, since the business technology relationship is fluid and continuous – not defined around a number of discrete "rules." This point is important for leaders to understand. In years past, especially when there was relative chaos in our technology organizations, we defaulted to sets of rules and regulations which more often than not were used to hit non-compliers over the head with. In our desire to create "order," we ended up offending our customers! In extreme cases, the business technology relationship was defined around a set of internal service level agreements written to trap clients into draconian consequences for less-than-egregious offenses. This kind of organizational authoritarianism is what explained the lack of business technology "alignment" for many years.

Reporting relationships should speak to business processes. CIOs and CTOs should report to the CEO or the COO, not the CFO, whose incentive is only to hold costs down. CIOs and CTOs – for as long as they exist – should organize around hardware, software **– and processes**. General Motors, for example, has recently identified five "process officers" responsible for key processes like supply chain efficiency and program management. Since GM is a decentralized organization, there are lines of business CIOs – who report to the lines of business CEOs or Presidents (with a dotted line to the enterprise CIO). The addition of the process officers is the enterprise's way of creating synergy across enterprise and line of business objectives.

Business technology leaders organize around business processes and – especially – objectives, not around self-contained best practices around, for example, data center management. In other words, data center management, desktop management, storage area network management, and the like, should all be organized purposefully, that is, with a clear connection to the business models and processes that they serve. When problems arise, solutions are sent through business filters first and foremost, and then filters that focus on other factors.

Leaders make certain that the computing and communications infrastructure works. This means that it's secure, reliable and scalable. This also means that it's cost-effective. Leaders understand that there are alternative ways to acquire, deploy and support computing and communications infrastructures. They optimize the alternatives with reference to their organization's core competencies, culture and evolving business strategy.

The ability to write diagnostic requests for proposals (RFPs) for infrastructure technologies and processes, as well as the ability to craft effective service level agreements (SLAs), are two indispensable leadership skills.

Another skill is measurement. How well is the infrastructure performing? What does the industry benchmarking data tell you? Leaders are aware of what's happening in their industry and in their environment, especially with acquisition trends. If outsourcing makes sense then lead the SWOT (strengths/weaknesses/opportunities/threats) analysis: the worst thing that could happen is for a tsunami of opinion about the desirability of outsourcing to overwhelm you, forcing you to react quickly (and probably badly). Leaders direct **all** of the technology acquisition discussions.

Leaders also manage their infrastructures cost-effectively. This is the commodity side of the business. The trends are clear here, so leadership will increasingly be about the acquisition **and measurement** of reliable, flexible and secure infrastructures.

Leaders communicate. They understand that the essence of communication (and its cousin, influence) are hard and soft facts, and hard and soft communications skills. Are you a good communicator? Do people understand what you say – and what you mean?

Communication is based upon the wide and deep understanding of your audience and your own specific strengths and weaknesses. Some people are natural communicators (or salespersons), while others find it hard to connect with colleagues or "customers." If you have natural talent here build upon it through practice, executive education and coaching. If you're hopelessly terrible at it, find someone on your team who's an effective communicator and give them the responsibility for your team's communications duties.

Communication is a continuous process. When things are relatively quiet, leaders still need to communicate what they're doing, the status of their projects and their strategies. When things are bad, they can call upon a deep continuous relationship with their partners and stakeholders to jointly solve problems. When things are good, leaders can exploit their communications investments to make sure everyone understands the significance of the victory at hand.

Leaders communicate good news, bad news and no news in a predictable, timely, digestible way.

Much communication is routine – about how well the infrastructure is performing, technology costs, technology total-cost-of-ownership (TCO) models, and the return on investment (ROI) of business technology projects and programs.

Project/program/portfolio "dashboards" are also a good idea: everyone likes easy-to-read status reports on key projects and whole programs. It's also a good idea to develop some form of "scorecard" that communicates the overall impact that business technology is having at the company.

Leaders think about who creates, distributes and maintains the technology "message" inside and outside of the company. Business technology leaders are sensitive to the need to internally and externally market their buiness technology accomplishments and strategies.

So what are the pieces of a good technology marketing strategy? First, consider what you're "selling." You're selling hardware, software, services, **image, perceptions and strategies**. When everything goes well, everyone thinks that the technology people are really pretty good, that things work reasonably well – and for a fair price. If the hardware and software works well, but the image is poor, technology is perceived to be a failure, just as bad hardware and software – but good perceptions – will buy you some time. Like everything else, you're selling hard **and** soft stuff, tangible **and** intangible assets and processes.

Next consider who you're selling to – noting from the outset that you're selling different things to very different people. Yes, you're selling hardware, software and services (along with image, perception and strategy) to everyone, but the relative importance of the pieces of your repertoire shifts as you move from office to office. Senior management really doesn't care about how cool the network is or how you've finally achieved the nirvana five-9s for the reliability of your infrastructure. They care about the 20% you lopped off the acquisition project you just launched, or how company data is finally talking to each other and that you're now able to cross-sell your products. The content **and form** of the message is important.

Public companies have a unique challenge. Increasingly, technology is included as a variable in company valuation models. This means that the analysts that cover public company stocks look at technology infrastructures, applications and best practices in order to determine how mature a company's technology acquisition, deployment and support strategies are. CIOs and CTOs may talk to these analysts, fielding their questions and otherwise molding their understanding of the role that technology plays in the current and anticipated business.

What's the brand of your technology organization? If you were a professional sports team, what would be a good name for your organization? Would you be the Innovators? The Terminators? Put another way, if you asked the analysts who cover your stock to word associate technology and your company, what would they say? Disciplined? Strategic? Weak? What about collateral materials? Does the technology organization have its own Web site? It's own brochures? Case studies? White papers? Referenceable "accounts" (internal customers who are happy with technology's services)? Are there newsletters and technology primers? Is there information about the competition?

Is there a technology "road show"? A consistent message about the role that technology plays in the company, how technology is organized, what matters most, the major projects, and technology's contribution to profitable growth – among other key messages – is essential to running technology like a business.

Most importantly, are there dedicated resources for technology marketing? I cannot emphasize enough the value of internal and external technology marketing. The technology story at your company – assuming that it's mostly good – needs to be packaged and sold on a continuous basis. Invest a little here and the payback will be substantial. Business technology leaders understand all this.

Why Is Everyone So Quiet?

The CEO ...
"So it's all my fault ... I've built a culture around good 'ol boys and gals – at least I have gals here ... do I get any credit for that?"

The General Counsel ...
"Be careful what you say ... this people stuff is hairy ..."

The President ...
"What did you say?"

The Chairman of the Board ...
"It's time the board got more involved here ... the changes in the management team were sold to us as the best thing for the company, but things don't appear to be working out ..."

The CLO ...
"We can fix this ... there's a lot of training we can do ..."

The CEO ...
"It seems we're past that ... it seems everyone wants more ..."

The CFO ...
"Who's 'everyone' ... this is a pretty small group ... are you saying that our people don't know enough about business technology convergence – or something else?"

The Chairman of the Board ...
"Yes ..."

The President...

"Let's level set: where are we now? How many of our people know the right things and perform the way they should? Do we know?"

The CLO...

"We could measure it ... it will take a couple of months ..."

The Chairman of the Board...

"Forget about it ... it's the wrong question anyway ... what we need to do is step back and think about our company, our culture, our core competencies and how we mix all that into profits ... OK?"

The CFO...

"OK."

The President...

"OK."

The CEO...

"I'll get ready to go ..."

The Chairman of the Board...

"That would be good ..."

The Facilitator...

"I think we're done here ..."

The Chairman of the Board...

"Thanks for your help ... I think we can take it from here ..."

Endnotes

[1] See "Meritocracy Vs. the Golf Culture," *Business Finance*, August 2002, for a great discussion about how the rules for advancement change the higher you go in the organization.

Chapter VIII

There's Just One More Thing …

I don't know about you but I'm exhausted.

We've been through a lot during just the past few years. Y2K, e-business frenzy, free capital, dot.bombs, corporate scandals, integration technology - and we're still standing. Of course we are. Because all of these "major" events, "disruptive technologies," and once-in-a-lifetime stories are anything but. Life goes on, as they say, in the trenches and in the clouds.

It's all about commitment. Business is not always complicated. We sometimes convert simplicity into complexity so we have something to do, into some problem that only we can solve. I guarantee you that if you change the way you think about business technology and practice sane management you will save money and become more competitive, more profitable. There is huge leverage here. Companies are wasting millions and in some cases billions of dollars a year because their business technology relationship is fractured.

Here are the questions:

- Do you think about business as heading toward collaboration?
- Can you identify the technologies that enable collaboration – and the collaborative business models that pull technology?
- Do you get the importance of technology integration?
- Will you organize things differently?

- Will you re-visit your business technology management best practices?
- Will you be a little more careful about who you invite inside your company?

How'd You Do?

Figure 41 cuts to the chase – and assumes answers to the above questions. It's also an oversimplification because very few important decisions are perfectly red, yellow or green. But it's still a good mantra.

Figure 42 presents the whole picture. Collaboration and integration mean specific things. Organization and management best practices can also be identified. If you invest in business technology with some reference to Figure 42, the business technology relationship will improve, you will reduce costs, and make money.

Figure 41. The investment filters

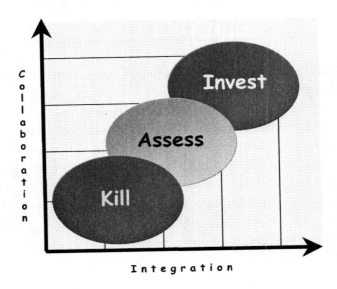

Figure 42. Collaboration, integration, organization and management

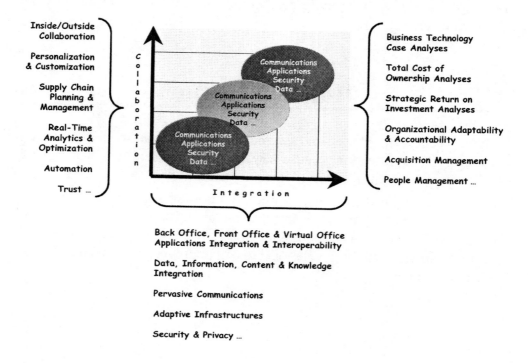

What's Not To Like About This Outcome?

If you followed the plot in the end-of-chapter conversations, you noticed that the CEO got whacked. Why? Looks like there were some major organizational, management and people screw-ups. The Chairman of the Board – who was quiet during most of the early conversations – decided to wake up and assert himself. Is this how it happens? Sometimes. But this is not about job preservation – though it's often a pleasant by-product of good business technology management. It's really about taking a hard look at a bad relationship that evolved into a very promising marriage.

There are lots more things we could review here, but we've talked enough. You've got the philosophy, the filters, and all of the lists. There's just one more thing I'd like to say: try to approach the business technology convergence challenge from a different place, not from an extrapolation of where you've been over the past five or ten years, but from a scenario of where it can take your business over the next five or ten.

The interplay between emerging collaborative business models and computing and communications technology is one of the most powerful drivers of the early 21st century. We talked a lot about the pieces. You can take it to the next level. Thanks for the time. I hope you think it was well spent.

About The Author

Stephen J. Andriole, Ph.D.

Steve Andriole's career has focused on the development, application and management of information technology and analytical methodology to complex business problems. These problems have been in government and industry. Dr. Andriole has addressed them from academia, government, his own consulting company, a global insurance and financial services company, and from the unique perspective of a venture capitalist.

Government

Dr. Andriole was the Director of the Cybernetics Technology Office of the Defense Advanced Research Projects Agency (DARPA) where he managed a $50M program of research and development that led to a number of important scientific and technological advances in the broad-based information, decision and computing sciences. While at DARPA, Dr. Andriole supported the development of spatial data management and multimedia systems, decision-support systems, computer-aided simulation and training systems, and intelligent technology-based command & control systems. Dr. Andriole's research program at DARPA seeded MIT's Architecture Machine Group, which evolved into the internationally-known Media Lab. The program also contributed to the development of the ARPANET (which became today's Internet & World Wide Web), interactive training simulations, such as SIMNET, and a whole host of artificial intelligence-based advances. Dr. Andriole has consulted with the National Academy of Sciences, the National Science Foundation, and various offices and agencies of the U.S. Department of Defense.

Industry

Dr. Andriole's career in industry has taken several forms. He's been a Chief Technology Officer at a Fortune 50 company, a CTO at a public venture capital company and private equity venture fund, and an entrepreneur.

Dr. Andriole was most recently the Chief Technology Officer and Senior Vice President of Safeguard Scientifics, Inc. where he was responsible for identifying technology trends, translating that insight into the Safeguard investment strategy, and leveraging trends analyses with the Safeguard partner companies to help them develop business and marketing strategies. Dr. Andriole was also a Principal at TL Ventures, the Philadelphia region's largest private equity fund. While at Safeguard, Dr. Andriole worked closely with many companies at all stages of their development. He was also the primary Wall Street contact for Safeguard, frequently interacting with the analysts that covered SFE.

Dr. Andriole was the Chief Technology Officer and Senior Vice President for Technology Strategy at CIGNA Corporation, a $20B global insurance and financial services company, where he was responsible for the enterprise information architecture, computing standards, the technology research & development program, and data security, as well as the overall alignment of enterprise information technology investments with CIGNA's multiple lines of business.

As an entrepreneur, Dr. Andriole founded International Information Systems (IIS), Inc., which designed interactive systems for a variety of corporate and government clients. IIS specialized in business requirements analysis and prototyping, the design of user-computer interfaces, and software systems evaluation. IIS also performed technology investment risk assessments for government and industry. He is also the founder of TechVestCo, a new economy consulting consortium that identifies and leverages technology trends to help clients optimize business technology investments, and the co-founder of Ascendigm, LLC.

Academia

Dr. Andriole is currently the Thomas G. Labrecque Professor of Business Technology at Villanova University where he teaches and directs applied research in business/technology alignment and pervasive computing. He is formerly a Professor of Information Systems & Electrical & Computer Engineering at Drexel University in Philadelphia, Pennsylvania, where he conducted applied research in information and software systems engineering, principally through the Center for Multidisciplinary Information Systems Engineering, which he founded & directed. The MISE Center generated over $5M in external funding, principally from industry, the federal government and foundations. While at Drexel, and as part of the MISE Center's R&D program, Dr. Andriole – with support from the Alfred P. Sloan Foundation – designed and implemented one of the nation's first totally online Master's program in information systems. This program continues today with some significant repeat industrial customers, including CIGNA and MetLife, as well as a large number of independent students. The program was featured in the nationally broadcast PBS special *net.learning*.

Dr. Andriole was a member of the faculty of George Mason University as a Professor and Chairman of the Department of Information Systems & Systems Engineering. He was awarded an endowed chair from the university becoming the university's first George Mason Institute Professor of Information Technology. The ISSE department was home to 30 full-time and part-time professors and had an annual external-sponsored research budget of $3M. Several research centers were established in the department including the Center for Computer Security and the Center for Software Engineering.

General

Dr. Andriole is a prolific author and speaker. Some of his 25 books include *Interactive Computer-Based Systems Design and Development* (Petrocelli Books, Inc., 1983), *Microcomputer Decision Support Systems* (QED Information Sciences, Inc., 1985), *Applications in Artificial Intelligence* (Petrocelli Books, Inc., 1986), *Information System Design Principles for the 90s* (AFCEA International Press, 1990), the *Sourcebook of Applied Artificial Intelligence* (McGraw-Hill, 1992), a (co-authored with Len Adelman) book on user interface technology for Lawrence Erlbaum Associates, Inc. entitled *Cognitive Systems Engineering* (1995) and a book for McGraw-Hill entitled *Managing Systems Requirements: Methods, Tools & Cases* (1996). He has recently published articles in *Software Development*, *IEEE Software* and *the Cutter IT Journal*. Overall, he has published over 100 articles and papers. He is also the author of a new 15-part series for *Datamation* on business/technology alignment. This series, which began in February 2001, examines how large, medium and small businesses can optimize their investments in information technology.

He is currently developing two new books: the first, *TechVesting: Leveraging Information Technology in the Digital World*, will describe a methodology for investing in information technology from three perspectives: creators of technology, users of technology and technology investors. This book will appear in 2005 and will be published by the Wharton Press/Prentice Hall Financial Times.

Some of his consulting clients have included General Electric, Magnavox, CIGNA, the Massachusetts Institute of Technology, Merck & Co., Computer Sciences Corporation, Air Products & Chemicals, GlaxoWelcome, the Software Productivity Consortium, and the Defense Advanced Research Projects Agency. He serves on a number of public and private corporate boards and on the Board of the Ben Franklin Technology Center of Southeastern Pennsylvania.

Dr. Andriole received his B.A. from LaSalle University in 1971 and his Masters and Doctorate degrees from the University of Maryland in 1973 and 1974. His masters and doctoral work was supported by a National Defense Education Act fellowship. His Ph.D. dissertation was funded by DARPA.

Details about Dr. Andriole's career can be found at www.andriole.com.

Index

Symbols

1st digital revolution 16
2nd digital revolution 16

A

Accenture 12
accounting systems 90
acquisition strategy 158
adaptive Infrastructures 111
advisory board 8, 16
agents 72
alignment 23
allocation 175
Amazon 72
analytics 96
anti-virus technology 119
Apple 24
application architecture 90
application integration 39
application migration 95
application programming interfaces (APIs) 94
application service providers 24
application technologies 97
applications 30, 84, 89, 104, 106, 168, 177
applications access 117
applications and communications architectures 114
applications architectures 224
applications Integration 24, 88
applications interoperability 88
applications portfolio 90
applications portfolio management system 32
applications service provider (ASP) 161
architecture 25, 39, 90, 128
Ariba 96
authentication 117
authorization 118
automated 55
automation 43, 53, 70, 91, 103, 106, 224
automobile industry 19

B

B2B exchange 86
back-office 32, 89
baseline assessment 147
BEA systems 91
bean counters 5
Berners-Lee, T. 73
biometric technology 119
Bluetooth 24
board accountability 9
board of directors 8, 16
broadband 85
broadband adoption 109
broadband wireless 110
budget 18

business analytics 33, 95, 225
business case 132, 144, 180
business convergence 75
business conversation 53
business intelligence 68, 93, 95
business models 30, 218
business process interfaces (BPIs) 63
business process outsourcing (BPO) 162
business resumption planning 117
business scenarios 53, 239
business skillsets 39
business technology acquisition 28, 226
business technology alignment 15
business technology assets 38
business technology cases 180
business technology convergence 28, 36, 56
business technology initiative 30
business technology management 20, 218
business technology optimization 38
business technology relationship 1
business-to-business (B2B) 23, 56, 64, 103, 163
business-to-business (B2B) transactions 35
business-to-consumer (B2C) 23, 56, 103, 163
business-to-employee (B2E) 56, 103
business-to-government (B2G) 56, 103

C

C-level executive 16
CD-ROM 21
cell phone 66
centralization 128
certificate authority technology 119
CFAR 157
chief alliance officers (CAOs) 10
chief executive officer (CEO) 1, 26
chief financial officer (CFO) 2
chief generalists 22
chief information officer (CIO) 2, 26
chief infrastructure officer 154
chief knowledge officer (CKOs) 10
chief learning officer (CLOs) 10
chief marketing officer (CMO) 7, 87
chief operating officer (COO) 1
chief partner officer (CPOs) 10
chief privacy officer (CPO) 2
chief security officer (CSO) 2
chief technology officer (CTO) 2
CIO Magazine 87
Cisco 12
claims processing systems 90
client/server 32
client/server computing 24
co-sourcing 159
COBOL 18, 94
collaboration 53, 90, 104, 129, 136, 224
collaboration officer 130
collaboration/integration 88
collaborative business models 58
collaborative models 76
collaborative planning 63
communications 84, 103
communications network 111
Compaq 29
competitive partnering 62
competitor intelligence 227
connectivity 59
consigliore 8
content management 24
continuous transaction 53
convenience 59
convergence 21, 56, 121, 141, 226
convergence conversation 15
core competency 229
corporate universities 229
cost allocation 171
cost structure 62
cost-effectiveness 19
cost/benefit 183
CPFR 63
CRM 68, 70, 86, 92
cross-selling 33, 54, 106
culture 133, 227
customer relationship management (CRM) 4, 20, 24, 69, 86
customer service 103
customers 67

customization 33, 53, 64, 91, 103, 224

D

dashboards 26, 68, 95
data 85, 104
data access 117
data center 113
data center operations 39
data communications 85
data integration 98
data mining 24, 33
data processing 24
data storage 99
data warehouse 24, 67
data warehousing 33, 69
database administration 33
database management and analysis 225
database management systems 90
decentralization 128
decentralized organizations 125
decision support 33
Dell 20, 58
demand forecasting 61
denial of service attacks 35
deployment 28
desktop 32, 89
desktop OLAP (DOLAP) 101
digital age 8
digital business-to-business (B2B) transaction 35
Digital Equipment Corporation (DEC) 21
digital revolution 15, 21
directory 113
discrete transaction 53
disruptive technology 17
distribution 65, 226
dot.com 15
dot.com bubble 15
dynamic pricing 56

E

e-business 18, 56, 69, 86
e-business application 69
e-business initiative 4

e-business strategy 32
e-learning 103
e-mail 58, 107
EAI 94, 97
eBay 72
electronic data interchange (EDI) 63, 103
Ellison, L. 115
encryption technology 119
enterprise application integration (EAI) 92
enterprise resource planning (ERP) 4, 18, 22, 63, 69, 86, 91
ethernet 36
ethics 228
ETL 102
executive conversation summary 15
extensible mark-up language (XML) 36, 63
extraction, transformation and loading (ETL) 94

F

faxes 107
fee-for-service 172, 175
fiber optic connectivity 108
firewall technology 119
Fortune 100 company 27
Fortune 500 company 22
front-office 32, 89
funding 143

G

Gartner Group 86
general counsel 2
generic structured knowledge 223
global positioning data 66
governance 128, 144, 155, 171, 188
grid computing 139
groupware 112

H

hard metrics 148
hardware and software incompatibilities 19
Harvard Business Review 76

HP 12, 29, 62, 157
HTML 92
human capital 218
human capital intelligence operation 233
hype-sters 7
hypertext mark-up language (HTML) 92

I

i2 63
IBM 25, 73, 85, 91, 154
IBM/PriceWaterHouseCoopers 12
IBM/SAP 156
in-sourcing 159
incentives 235
industry specific knowledge 224
information economics (IE) 181
information technology (IT) 18
information warfare 35
infrastructure 31, 68, 71, 84, 128, 168
infrastructure funding 175
infrastructure management technology 115
innovation 125, 137
inside collaboration 58
integration 20, 23, 27, 53, 90, 136, 157, 225
intelligent agents 20, 71
intelligent systems technology 74
interactive marketing 24
internal rates of return (IRR) 181
Internet 20, 23, 55, 57, 103
Internet bubble 20
Internet protocol (IP) 85, 111, 116
Internet/Web applications 32
internship programs 233
interoperability 20, 88
intranets 58
inventory control systems 90
IP (Internet protocol) 85, 111, 116

J

Java 24, 36, 97

K

key risk 183
killer app 57
knowledge management 24, 100
knowledge storage 100
KPMG 12

L

laptop 32, 67, 85
life cycle 67
Linux 24, 156
LotusNotes 104

M

m-commerce 109
Macintosh 24
mainframe 32
management 28, 103
management practices 57
manufacturing 226
manugistics 63
marketing 65
mass advertising 64
mass customization 56, 64
mass marketing 64
measurement 143
mentoring 235
meritocracy 218
messaging 112, 224
messaging applications 111
metadata 33
metrics 42, 225
Microsoft 12, 25, 85, 97, 154
Microsoft Exchange 105
Microsoft Office 36, 157
middleware 94
minicomputers 111
multi-tier applications 24
multidimensional OLAP (MOLAP) 101

N

National Institute of Standards and Technology 87
natural language understanding 43
net present value (NPV) 181
network access 117

network centric applications 39
network services 24
networks 104
newspapers 67
non-shared contracting 161
Novell 157

O

object data architecture 100
off-shore outsourcing 163
OLAP 101
OLTP 101
online analytical processing 33, 69, 100
online exchange 24
online exchanges 24
online transaction processing 100
operations officer 132
optimization 53, 60, 103, 224
Oracle 25, 85, 91, 154
organizational leadership 243
organizational model 130
organizational structure 125, 130, 189, 218
outside collaboration 58
outsourcing 132, 143, 158, 160
overhead 169

P

pagers 85
Palm 12
participants 1
partner management 103, 225
PCs 111
PeopleSoft 12
Perl 24
personal computers (PCs) 6
personal digital assistant (PDA) 12, 32, 66, 85, 105
personalization 33, 53, 64, 91, 103
pervasive communications 103
portals 93
president 1
PricewaterhouseCoopers 87
privacy 24, 34, 65, 84, 117
privacy compliance technology 119
privacy officer 2
privacy services 113
process improvement 26
process officer 132
process organization 132
professional communications 226
program management 39
project and program management 225
project management 39, 184
propeller head quotient 6
propeller heads 6
protection 8
public companies 4

Q

query 113

R

rapid economic justification (REJ) 181
re-engineering 127
real options valuation (ROV) 181
real-time analysis 106
real-time Analytics 53, 68, 103
real-time computing 139
real-time optimization 90
real-time transactions 101
Red Hat 157
regulatory trends 226
relational OLAP (ROLAP) 101
requests for proposals (RFPs) 160, 244
return on investment (ROI) 19, 23, 69, 87, 128, 132, 143, 245
return-on-investment (ROI) 23, 69, 87, 132
risk assessments 117
routers 111

S

sabotage 35
sales 65
sales force automation (SFA) 70, 86
SAP 12, 92, 156
scenario development 42
scenarios 77, 227
security 34, 75, 85, 113, 225
security requirements 24
security solutions outsourcing 163

semantic Web 43
service 226
service level agreements (SLAs)
    162, 244
SFA 71
shared risk 161
shareholder value 13
shop floor manufacturing systems 90
Siebel 12, 92
skillsets 39
SLAs 162
small office/home office (SOHO) 103
SOAP (simple object access protocol)
    97
soft metrics 148
spam 35
standard applications architecture 90
standardization 105, 143, 152
standards 20, 63, 97, 128, 153
Strassmann, P. 23
strategic planning 30, 42
strategy officer 132
supplier integration 103
supplier management 61
suppliers 11, 32, 58
supply chain 69
supply chain integration 21
supply chain management (SCM) 53,
    60, 70, 91, 157, 224
supply chain planning
    11, 20, 56, 60, 103
supply chains 55
support officer 130
SuSE 157
switches 111
SWOT (strengths/weaknesses/opportu-
    nities/threats) 244
Sybase 101, 154
synchronized 112

T

taxation 172
TCP/IP 77
techies 6
technologists 2
technology acquisition 28
technology bunkers 6

technology conversation 84
technology facilitators 26
technology infrastructure 30
technology integration 78, 129
technology investments 54
technology optimization 23
technology organization 16, 17
technology organizations 127
technology skillsets 39
technology standards 61
technology utilities 163
teleconferencing 103
thin client applications 32
thin clients 115
thinfrastructure 115
timing 163
total cost of ownership (TCO) 23
total economic impact (TEI) 181
total quality management (TQM) 184
total quality software management 26
total-cost-of-ownership (TCO)
    87, 132, 143, 245
transaction processing 103
trust 34, 74, 103
turf conversation 125

U

uber-filters 53
UDDI 97
unified communications 105
unified messaging 107
universal data access (UDA) 33, 101
Unix 156
up-selling 33, 106

V

value analysis 179
value chain 61
variation 36
vendor relationships 233
vendors 27
venture capitalist 11
virtual private networks (VPNs) 24, 63
virtual-office 32, 89
virus 35, 75
voice 85
Voice Over IP (VoIP) 111

voice recognition 72
voicemail 107

W

Wall Street analyst 2
Web 21
Web Services 20, 42, 85, 92
whole customer management 55, 67
Windows 156
wireless 43, 85
wireless collaboration 110
wireless communication 108
wireless networks 139
wireless technology 109
word-processing systems 90
worker bees 2, 11
workflow 58, 112
World Wide Web 17, 24, 56, 103
World Wide Web Consortium (W3C) 74
WSDL (Web Services description language) 97

X

XML (extensible mark-up language) 97, 157

Y

Y2K 17
Y2K compliance 23
Year 2000 17
Year 2000 compliance 88
Year 2000 problem 23

New Releases from Idea Group Reference

Idea Group REFERENCE

The Premier Reference Source for Information Science and Technology Research

ENCYCLOPEDIA OF DATA WAREHOUSING AND MINING

Edited by: John Wang, Montclair State University, USA

Two-Volume Set • April 2005 • 1700 pp
ISBN: 1-59140-557-2; US $495.00 h/c
Pre-Publication Price: US $425.00*
*Pre-pub price is good through one month after the publication date

- Provides a comprehensive, critical and descriptive examination of concepts, issues, trends, and challenges in this rapidly expanding field of data warehousing and mining
- A single source of knowledge and latest discoveries in the field, consisting of more than 350 contributors from 32 countries
- Offers in-depth coverage of evolutions, theories, methodologies, functionalities, and applications of DWM in such interdisciplinary industries as healthcare informatics, artificial intelligence, financial modeling, and applied statistics
- Supplies over 1,300 terms and definitions, and more than 3,200 references

ENCYCLOPEDIA OF DISTANCE LEARNING

Four-Volume Set • April 2005 • 2500+ pp
ISBN: 1-59140-555-6; US $995.00 h/c
Pre-Pub Price: US $850.00*
*Pre-pub price is good through one month after the publication date

- More than 450 international contributors provide extensive coverage of topics such as workforce training, accessing education, digital divide, and the evolution of distance and online education into a multibillion dollar enterprise
- Offers over 3,000 terms and definitions and more than 6,000 references in the field of distance learning
- Excellent source of comprehensive knowledge and literature on the topic of distance learning programs
- Provides the most comprehensive coverage of the issues, concepts, trends, and technologies of distance learning

ENCYCLOPEDIA OF INFORMATION SCIENCE AND TECHNOLOGY
AVAILABLE NOW!

Five-Volume Set • January 2005 • 3807 pp
ISBN: 1-59140-553-X; US $1125.00 h/c

ENCYCLOPEDIA OF DATABASE TECHNOLOGIES AND APPLICATIONS

April 2005 • 650 pp
ISBN: 1-59140-560-2; US $275.00 h/c
Pre-Publication Price: US $235.00*
*Pre-publication price good through one month after publication date

ENCYCLOPEDIA OF MULTIMEDIA TECHNOLOGY AND NETWORKING

April 2005 • 650 pp
ISBN: 1-59140-561-0; US $275.00 h/c
Pre-Publication Price: US $235.00*
*Pre-pub price is good through one month after publication date

www.idea-group-ref.com

Idea Group Reference is pleased to offer complimentary access to the electronic version for the life of edition when your library purchases a print copy of an encyclopedia

For a complete catalog of our new & upcoming encyclopedias, please contact:
701 E. Chocolate Ave., Suite 200 • Hershey PA 17033, USA • 1-866-342-6657 (toll free) • cust@idea-group.com